Viridian Landscape Architecture
PO Box 389
Pacific Grove, CA 93950

竹
大

The Friendly

Bamboo Revolution

The Friendly

Bamboo Revolution

by

Lawrence Stanley

Happy Shooting!

L. Stanley

Published by Bamboo4Sale.com in 2007
Printed in California

Dedication and Acknowledgments

This book is dedicated to my family: to my excellent wife Lisa and our ten children, David, Rachel, the twins Julie and Jonathan, Rebecca, Daniel, Roger, Anna, Robert and Andrew, who know all about living in the midst of a growing bamboo landscape.

A special thanks to my mother who broke out of retirement to edit and proofread this manuscript.

I am happy to acknowledge the staff and management of Bamboo Giant Nursery, who worked hard contacting customers and arranging for interviews and site visits, and also the talented crew who care for and install the bamboo. I also extend thanks to the volunteers of the American Bamboo Society and the European Bamboo Society who organize annual meetings and tours that have been very helpful to me in preparing this book.

Special thanks also to many friends in England, France, Germany and Italy who spent valuable time teaching me how to propagate and sell bamboo, and friends in China and Vietnam who included me in the production of bamboo for product manufacturing.

Palmata leaves in the morning.

Contents

The Friendly Bamboo Revolution

This book is about using bamboo in the landscape, both commercial and residential. Although we will touch on other uses for bamboo such as crafts and as a great food source, this is not intended to be a comprehensive reference book, a scientific journal or an exhaustive tutorial. Instead we will enjoy a romp through the brave new world of bamboo as an alternative to traditional landscape and an exciting new element in design. Many examples are drawn from our years of operating one of the largest bamboo growing and installation companies on the West Coast of America. Others are drawn from extensive visits to all the native homelands around the world of the bamboos we love. Unfortunately, some of my favorite bamboo books have a distinct Australian or English point of reference and their experiences do not directly relate to our markets or climate. This book is based on our experience throughout the West Coast of America where we see a dramatic range in climate, conditions and design goals.

Introduction - Why Bamboo?

It is more than a plant; it is a new design force. It is a revolutionary new material that is globally very old. It is a friendly, warm, but somehow profound and spiritual plant. It is bamboo. Just the word alone has tremendous association. It is one of those key words that can sell a product or design just because it is bamboo. When I am asked what I do and I answer, "Grow bamboo", it almost universally brings a smile and a twinkle in the eye. Then inevitably I hear about their latest experience with bamboo. "Oh, we just put bamboo flooring in our kitchen. We love it. It is so warm and beautiful, yet exotic." or "Wow, when we were in Hawaii we visited a bamboo forest. It was amazing. I had no idea. As we walked through the towering canes, with the wind whispering above, it was like being transported to another world." or "My neighbor has bamboo and it keeps coming up in my yard. I hack it down but it keeps coming back. I hate that stuff." In one way or another, bamboo brings a reaction.

Designers the world over have found that

Giant vivax bamboo culms - one of the running timber bamboo.

Bamboo furniture, mats, fencing, art, clothing, dishware and birdhouses - you name it and the "all bamboo" web store Bamboo4sale.com sells it. They carry hundreds of bamboo items produced in Asia by Symphony Bamboo Products. Shown here is the Symphony booth at a Lawn and Garden show held in Alameda, California. Bamboo is HOT!

bamboo sells. Bamboo clothing, mats, furniture, fencing, as well as bamboo flooring, cutting boards and household items are all hot. Bamboo shoots are imported by the ton. Bamboo hedges are the perfect solution for instant privacy. But we also find products that are not bamboo being sold successfully. Lucky bamboo is everywhere, the perfect little plant around the house or office. Just stick a cutting in water and it's happy, but it is not really a bamboo. Heavenly bamboo was considered a big hit in the yard a few years ago until people found out how quickly it gets out of hand. It's also not a bamboo. Restaurants like "The Bamboo Garden" or the fast-food chain "Bamboo Express" do not actually serve bamboo shoots.

Bamboo has been a key element of design in Asia for centuries. Europe has been "bamboo crazy" for years. In North America we are just getting the "bug". It is not a fad, but a real trend with solid value and lasting utility. One of the best ways to use and enjoy bamboo is to use bamboo elements in the landscape.

I have seen many very poor uses of bamboo, where bamboo is planted simply because the owner loved bamboo and wanted it everywhere. Bamboo should be selected and used as part of an overall landscape plan that follows the age-old landscape design principles. We must not forget the relationship between shape, line and proportion. Good design is still good design. We must still consider the elements of the site and the style desired and then use the diverse texture, mass and color of bamboo to create that pleasing design.

A small retail shop in China selling bamboo carvings naturally makes use of bamboo shelving and adds a nice touch with a bamboo hanging sculpture. Bamboo is a historical and cultural part of all Asian countries.

Bamboo Express in Cabo San Lucas, Mexico, is a Chinese fast-food chain which does not even serve bamboo shoots. Just the word "bamboo" works by association to attract the consumer.

BAMBOO NAMES
Because our plant-naming system is based on the flowers of plants, with bamboo's infrequent flowering, its names are quite irregular. Rarely does a year pass when a few are not re-classified. To keep my life simple, I use the common name for each bamboo. I created a common or short name from the Latin for any bamboo lacking a common name and use this name consistently throughout the book. Where clarity is important the Latin name is in (). The indexes in the back of the book provide a cross-reference. For example, on the previous page,
     vivax bamboo (*Phyllostachys vivax*)
shows my short name vivax for this bamboo. A common or short name greatly helps landscapers and customers not familiar with bamboo to get comfortable with it more quickly.

Design Principles

As always in designing a landscape, we must think of abstract shapes. Proportion is fundamental as it is in any kind of design, but in the landscape we are combining static objects and growing plants, a significantly more challenging task. Bamboo's unique growth pattern will take some re-training for the experienced designer and may actually be easier for the beginner who will not have to break any old habits. Balance is achieved when the positive masses such as plants and structures are in harmony with the negative spaces such as lawns and paving.

I like to start with those elements that are fixed by the project. The house, property boundaries, terrain, and views outside your space all must be either used or hidden. Next consider that landscape should enhance a lifestyle, not get in its way. For whom are we designing, what will they do with the space? How do they live, and what is important to them? What is the time line, and how long do we have to achieve the goal? These are often overlooked aspects that are critical.

Time Line

The time line is the biggest challenge I see in many current landscape designs, and one of the primary reasons clients come to our nursery. In much of urban North America today, the time window is very short. Many clients have a one-to-three year window in which they expect to create and enjoy the landscape. Then resale value is the primary consideration. Planting for the long term does not match the lifestyle of the upwardly mobile. If the yard does not have a shade tree, there are two options: either move a large tree in or find another solution for shade such as a structure. Planting an oak tree and waiting 15 years for it to grow up is just not practical. The other great failing of many designs is that in the rush to create the final design quickly, plant material is used that grows very quickly, but unfortunately does not stop growing. Although the current owner may move on, the house remains. Before long, the landscape has grown completely out of proportion and is beyond control so must be destroyed.

In working on budgeted landscape projects, the usual reaction when an attractive design is proposed that is over budget, is to scale down the plant material. "Well, let's just change those 15-gallon plants to 5-gallon size". This is the worst possible solution with bamboo. A better approach is to revisit the concept and try to build an equally balanced space with fewer elements. Keep those elements full-sized so the project makes a real impression on the first day.

Bamboo's growth pattern offers some very unique properties that can be used to create landscape elements full-size that do not get any larger over the years. Bamboo can also spread and fill in quickly and cost-effectively.

The Space

In exploring the lifestyle and goals of the client, we usually find several key places and activities that are the focal points about which the design must revolve. Most people, in reality, spend very little time in the landscape. A realistic assessment will allow the design to be optimized for the lifestyle. It is refreshing when a client will admit that he really plans to spend little time in the yard and simply wants to minimize yard work. We then focus on the views from the windows or street. For another client, the yard may be used for the occasional big event. Others may focus on concealing a hot tub or patio where they plan to spend considerable time. For most, the front yard lacks curb appeal so we consider the view from the street or driveway. Once the key points have been identified we begin with some rough sketches from those critical viewpoints. The fixed elements are put in and then we start decorating, placing shapes like furniture with shape and depth.

Shapes

As we begin exploring the shapes we want to use in the landscape, we first consider the surrounding landscape. Do we have a view? Do we see mountains, trees or the beach? Can we echo these shapes in the landscape to unite our small space

Some Handy Design Tips

The goal is to capture the essence of the site and the design concept for two purposes. First, so you can see if the vision you see in your mind is really what you want, and second, so you can convey the concept to someone else. The method should be fast and expressive.

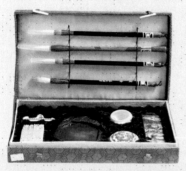

Some designers can use a pencil and paper, others pen and ink and some rely entirely on their computers. I find that I can work very fast with ink and brush, not exactly in the Chinese style, which involves making your own ink and following tradition, but with the same expressive quality of the brush and minimalist portrayal of shape and texture.

1 I use a watercolor brush that is intended to have a water reserve in the handle. I instead fill it with ink. With it I can make lightning-fast sketches. I also use a few basic water colors with a large brush to add color washes. It is all portable and clean; just snap the lid closed when done. The sketch is dry in minutes and the brush rinses with water. The sketch featured here took less than 15 minutes.

6 So now I have a sketch, and it is time to make it real. I found plenty of good pictures of Torii by a web search, so I printed out a picture and measured the proportions. I poured concrete posts and built a redwood frame. From my sketch I knew I wanted a berm covered with low bamboo with a fine texture and light color to hide the driveway. White stripe is perfect, so I poured a few raised concrete pads a foot above the ground for my pots, so that once the white stripe fills in, the pots will be visible. Bright orange is the traditional color which contrasts nicely with black trim, the black rim of the gong and black pots. This was a two-weekend project and by next year the white stripe bamboo should be giving good coverage. By year two, the project will be perfect.

2 I started by sketching the existing scene with my brush and ink. The prime view of the koi pond had a great backdrop of moso bamboo with a stone path close to the pond. My driveway was in between. I clearly lacked a focal point, and the driveway needed to disappear. But what to do?

Here is where technology comes in. By copying the sketch at this stage I can rough in all kinds of ideas until one looks right.

3 I had tried a couple of red maple trees, some large rocks, a few benches, and then a Torii. Perfect. The Torii is commonly seen in Japan marking the entrance to temples, separating the space. This point separates the serenity of the koi pond from the bustle of my driveway.

4 Something was missing. I needed an accent on each side. Photoshop is also a handy tool. I sketched some options and tried them in the spot. The round black pots with the long floppy dark green leaves of tessellatus bamboo was right.

5 Now I added a few details, a gong hanging from the Torii, my plants and a little color. It looks good.

with the larger space? Even one stunning tree or rock can create the theme for the whole design.

If our surroundings are not natural, then we may have walls, buildings and other urban features to work with. We can match or contrast. These scenes are defined by hard lines, sharp edges and repeating patterns, quite unlike the random softer shapes of nature. We can also build with more defined shapes.

Shapes can also be drawn from those fixed elements like the house, pool and any feature "off limits" for change. Scale now becomes a key aspect of the design. The elements of the landscape should be on the same scale as these fixed features.

Texture

Perception is not an exact process. Our impressions are very much dictated by the mental processing that accompanies our sense of sight. Converging lines indicate depth. Stripes can make an object appear tall or wide. When a painter blocks in the elements of a new painting, the first test is to cover your eyes, look toward the painting and uncover the eyes. What do you see first? What path does the eye follow? Where does it end? He wants to capture your view, and lead you through the elements of the painting to a final destination. The landscape scene also needs to accomplish this. The eye should not be jumping from one detail to another, but should be smoothly taken through the scene and focused on the key element. Only a strong three-dimensional design will accomplish this. If the space is limited, create the feel of space with geometry and texture. If the telephone pole next door is a major distraction, hide it with a tall bamboo.

Style

Style has two important impacts on the design. The primary role of style now comes into play as we begin interpreting shapes and textures into real objects such as plants, stone or plastic. But style also has an influence in that first conceptual exercise with the space and shapes used. Historically, style was derived from the site, as most landscaping originated on or near the property. The country style integrated the rural setting with the landscape. In today's urban setting we have much more freedom

The modern style is characterized by contrasting textures. Mexican weeping bamboo above has long narrow leaves but they hang down forming a rounded shape. Linearis bamboo below also has long narrow leaves, but they are sharp and upright. The two bamboos are complimentary yet in stark contrast. They have simple clean lines and fit well in a modern design.

in creating styles from other regions or cultures.

Country

Wood, stone and a random unkempt aspect suit the country style particularly well in a setting with significant external natural features. An exuberant proliferation of growth is typical. A less well-behaved bamboo such as yamadorii with its mixture of green, red and yellow canes and floppy leaves would fit this style.

Modern

Simplicity of line and functionality of space dominate the modern style. Functional elements such as hot tubs and swimming pools are focal

Yamadorii bamboo just seems to fit the country style. Notice the leaves are not well ordered yet its amazing vigor gives it the zest required for this natural style.

points.   Straight vertical privacy hedges are equally functional yet fit this style perfectly.   Strong lines require contrasting textures. Mexican weeping bamboo with its long softly hanging leaves can be contrasted with strongly vertical culms and upward pointing leaves of linearis bamboo.

Formal

There is a classic elegance to the formal style, where man's mastery over nature is displayed.   Symmetry and balance mandate well-behaved plants.

Shiroshima with its large brilliantly variegated leaves is very attractive in the Mediterranean-style setting, especially with a masonry backdrop.

This generally means a tremendous amount of maintenance, as hedges are sculpted and formed against the nature of the plants used. Here is another opportunity to use the unique nature of bamboo to your advantage. Trimming and sculpting is only required once, with just the annual elimination of new shoots. Hedges made from albostriata with its boldly variegated leaves or humilis with its plain green will not look like bamboo at all, and can be used to simulate a bay hedge. Chinese goddess bamboo would look like a box or yew hedge, but without the scalped look these hedges get over time.

Colonial

The colonial landscape is more comfortable than the formal style. It is still clean and neat but with more natural materials and random appearance. Well-behaved bamboo growing in containers without trimming would fit well. Several large stands of timber bamboo with lawn between would reinforce the self-confidence of the style.

Mediterranean

Outdoor living characterizes the Mediterranean style. Plants support that function by providing shade or attractive backdrops. The brilliantly variegated bamboos such as shiroshima fit extremely well in this setting. Small potted bamboos spilling over

the side, and gracefully arched culms of a borinda bamboo are almost excessively rich in contrast to the sun-baked earth tones of stucco and paving.

Chinese goddess has very small tight leaves in organized rows which can be trimmed into a very neat package. It looks particularly good in a shady location in a formal garden.

Veitchii is a favorite in Japanese-style gardens. The margin of the leaf dies creating the unique tan color around the edge.

Oriental

This is the least attractive setting for wildly growing bamboo. The oriental style is characterized by controlled calm and meticulous care, with a great sense of age. Here is the place where bamboo is controlled, contained and able to provide the design elements needed without all the maintenance that such a design seems to indicate. Bamboo such as veitchii is prized for its improbable precise tan margin on the edge of each leaf. Square bamboo, upright with long drooping leaves, is another favorite.

Creating Architectural Forms

We have discussed overall landscape styles, but the way plants are used can create a specific element with a defined purpose. We will briefly explore these elements and how plant selection and planting methods can create specific design elements. The distinct bamboo culms can be used to lead the eye or create a vertical texture. The brilliant color of some culms which remain year-round can very favorably replace the splash of color created by flowering plants. A number of bamboos have a tropical look with very large hanging leaves, yet they are cold hardy. They create a complementary sculpture to a man-made feature.

Plant selection can be simplified by considering a number of planting categories based on function. These categories are defined by the role the plant will play in the landscape, not by the plant itself. Bamboo can be used in all these categories depending on the scale and style of the particular project.

Specimens

These plants form a focal point in the landscape. There may be one such

Borinda is a magnificent clumping bamboo with a luxuriant crop of small leaf clusters. It is inviting and perfect in a Mediterranean style setting.

Tibetan princess is a specimen in almost any setting. Here, hanging over the koi pond it steals the show. It is a clumping bamboo that does well in sun or shade.

point or several, viewed from different vantage points. They are distinct in color or form from the background so they are clearly delineated. Bamboo lends itself to this role with relish. A large specimen of Tibetan princess bamboo was used to make a focal point on the far side of a pond. This tight clumper formed densely foliated culms that arched out over the water. The bluish-green color stood out against the pale green background created by a moso bamboo grove. In another case, a clump of Robert Young, a yellow culmed running bamboo with 2-inch diameter canes and about 25 feet tall was placed in a large pot in a garden setting with shorter round-shaped bamboos. When the morning sun caught the brilliant yellow culm, the view out of the bedroom window was spectacular. Trimming the lower branches from bamboo and revealing the culms is a great way to convert another green plant into an amazing specimen.

Backdrops

Just as the photographer can roll down a backdrop to best set off the subject, so we plant backdrops. They must create a permanent evergreen framework for the rest of the design. Again this plant material is usually the farthest from the subject and involves the coverage of the largest area – usually at the perimeter. It is therefore the most sensitive to the size of the planting and to the budget. The bamboo privacy hedge is superb in this roll. The spreading nature of bamboo allows this screen to fill in quickly yet even the initial planting will give significant coverage. Bamboo species are selected for this role based on desired height and neutral green coloring with a fine texture that blends the plants together, creating the uniform impression of a vast bamboo forest, not a 3-foot-wide hedge.

Backdrops can also be a low ground cover, such as white stripe

Palmata makes a great garden accent. The large leaves at waist height are truly exotic.

The bamboo hedge in the background offers just the right backdrop for this decorative bamboo fence. It is soft and rounded in stark contrast to the lines and angles of the bamboo construction.

bamboo which reaches a maximum height of about 1.5 feet tall, and can be mowed down each year to keep it vibrant. It will control weeds and makes a stunning bed on which to place a statuary, rock or fountain.

Accents

This final category usually includes the flowering plants which add a seasonal splash of color. But the same interest can be achieved with bamboo that exhibits unusual coloring or form. Palmata with its large leaves clustered like fingers on a hand at the end of 4- or 5-foot tall branchless culms is striking and unique and adds the necessary interest of this category.

Permaculture

Permaculture is a design system for creating sustainable human environments. Although it is all about plants, animals, building and energy, it is more about the relationships between them. In short, permaculture blends the structures and landscaping requirements with the inherent qualities of both animals and plants to create a life-supporting system that is both ecologically and economically sound in the smallest space. When fully implemented, these principles lead to total self-sufficiency. In less extreme cases however, the principles allow the most to be made from the landscaping choices. Under these guidelines, function takes precedence over form. Each plant added to the landscape needs to serve a function, and give back more than it takes.

Bamboo can play a key role in creating a habitat that is in harmony with nature. For centuries, bamboo has been a key component of living in Asia. It provided shelter, food, tools and art. As we now evaluate bamboo in our society, we find the culms to be a great source of poles for building structures, garden stakes, fencing, furniture building and concrete reinforcement.

The living plants are extremely effective for erosion control, as windbreaks and as a source of both human food and animal forage. Bamboo is particularly interesting as animal forage as it stores a much higher percent protein in its leaves during the winter when other grasses-stored protein are at a minimum. Because it operates on a cycle just opposite other forage grasses, a planting of bamboo in a hedge along the edge of a pasture can beautifully supplement the conventional pasture.

Bamboo is harvested each year sustainably. It is used as a low-impact building material. Here canes are harvested and split with hand tools and used to make a roof and the cabins above.

One-third of the bamboo biomass can be harvested each year in a sustainable way. No other tree can provide such an abundance, and so quickly after planting. Bamboo also provides opportunities to be planted in conjunction with other plants without root competition. Bamboo is a tall tree-like structure with a very shallow root system allowing it to share space with deeper-rooted species.

Up to a third of the bamboo culms can be harvested for poles each year without damaging the bamboo grove. Just a small clump in each yard provides all the bamboo products needed for the family.

Bamboo Selection Criteria

Bamboo is the epitome of an environmentally sustainable plant that is used to produce green, safe products and upscale high-performance products. It is a plant with spiritual connections and with deep historical roots. It is also a cost-effective solution to many landscaping problems, but cost is not driving the North American market. Instead it is seen as a premium product. Bamboo has an exotic association because of its unique features. In this section we will examine the six primary attributes of bamboo which are relevant in selecting a bamboo for the landscape. Understanding bamboo will allow the design elements identified in our conceptual sketches of shape, texture and style to be realized in the right bamboo species.

The Leaves

Asian bamboo art reduces the essence of bamboo to its simplest components. A few brush strokes of black ink on white paper and it is clearly bamboo. The two elements are the nodes and the leaves. The leaves do not form as tree leaves form. They are actually extended leaf sheaths which are wrapped

These borinda leaves hang with the classic look.

around the branches. They are longer than wide, with a rounded base and pointed tip. A vein runs down the center. They are glossy on top and matt below. It is the bamboo leaf that brings sound and movement to the bamboo plant. Its small leaves rustle in even the lightest wind making a relaxing music for tired ears. The sound is distinct just as a poplar tree makes a distinct sound, but the bamboo is more mellow, like a thousand whispering voices. The leaf is the essence of bamboo and must be considered carefully when designing the

Bamboo leaves are extended leaf sheaths as shown on this variegated albostriata.

The palmata leaf shows off the structure particularly well.

landscape. When a new customer focuses first on the color of the culm or the height of the bamboo, I always point to the leaves instead. So we start this examination by considering the leaves.

Evergreen and Tough

The most amazing feature of the bamboo leaf is that it stays green all year-round. When you think of your common evergreen options, bamboo stands out in stark contrast as being much more like a deciduous tree in form and in its ability to screen. Unlike the tough, almost plastic quality of most evergreens, the bamboo leaf is thin, light yet very tough. Rather than being rigidly attached, it has a flexible connection, showing motion with even the smallest breeze. Yet for all its flexibility it is well attached and hangs on in the worst of the winter storms.

Bamboo can handle abuse unlike most landscaping plants. It can survive total defoliation due to extreme abuse or very high winds and will re-leaf in a matter of weeks. A 35-foot tall bamboo with a large spread in a 5-foot diameter wood planter was moved into a parking lot to give some shade. Unfortunately, it was not in an area covered by the irrigation system. For weeks it was quite forgotten until the leaves turned yellow and started dropping. It was thought to be dead, but was manually watered anyway. Within two weeks new leaves appeared and it was as good as new. In another planting, a tall bamboo not particularly well-suited for high winds was planted in the yard of a house located right on the top of a cliff overlooking the northern California beach. A brutal winter storm shredded the leaves and stripped the plant. It also re-leafed.

Color and Size

Bamboo leaves range dramatically in size. Tessellatus leaves in the foreground just dwarf the small leaves on the giant timber bamboo Robert Young.

Bamboo leaves vary by species in both size and color. The larger timber-type bamboo has leaves ranging from 3 inches long by ½ inch wide to 6 inches long by 1 inch wide. Color ranges from a pale olive-green to a rich dark forest green. All turn yellow before dropping. Some are upright and others hang. Medium-size and smaller bamboo have a great range of leaves from long wispy streamers to huge tropical-looking leaves with a range of shape and habit. All bamboo leaves are green or variegated with white, cream or yellow.

Bamboo will drop leaves throughout the year but have a heavier than normal drop in the spring when the plant is producing new shoots. As leaves survive more than a year, leaf drop is less than half that experienced with deciduous trees, and the small leaves make attractive mulch. Bamboo

leaves contain silica which is needed by the plants, so allowing the leaves to collect and decompose is a good thing.

The bamboo leaves really set the tone of the plant with different leaves appealing to different personalities, and fitting different landscape goals. Some of the largest bamboo such as moso has delicate fine pale leaves that form layers that seem soft and relaxing. Others like vivax have bold upright leaves that seem like soldiers lined up ready for action. Yet others like the *Bambusa* family have long straggling leaves in disarray, sloppy to some but disarming and comfortable to others. No other plant offers such a range of personality, and when a perfect match is made, no plant bonds so well with its owner. Once a customer brought in two beautiful bamboo plants he had purchased several years before. He explained he had been transferred and was moving but could not take his babies. He wanted to give them back because he wanted them given a good home and cared for properly. It was a tearful parting. Bamboo is like that.

Vivax is a giant timber bamboo of the running type. Its ordered long leaves give a formal impression.

Oldhamii, one of the *Bambusa* family, is a giant timber bamboo of the clumping type. Its floppy leaves give a very casual impression. Compare it with the other giant timber bamboos on this page.

Moso is a giant timber bamboo of the running type. Its small delicate leaves give a soft, inviting impression.

This hillside in China of giant moso bamboo forms layers and a sharp contrast that is clearly visible even at a great distance. This year-round landscape is unmistakably bamboo.

Plant Status Indicators

Bamboo leaves also tell the gardener how the plant is faring. The first sign of trouble is the longitudinal curling of the leaf. This means that the plant is not getting enough water up from the roots, which has two possible causes. The ground may just be too dry, or the ground may be two wet. Bamboo roots require air to extract and pump water, and saturated ground is as much a menace as dry ground. Push your fingers down into the earth under the plant. If it is dry, then the plant needs a heavier water schedule, and if it is wet, then either cut back on the watering or better yet, provide adequate drainage. This can be accomplished with either actual drainage ditches or by mounding up the ground so the plant is higher above the water table. Not only will this make the plant very happy, but the saturated ground around it will keep it from spreading (see containment section).

Sustained heat or drought or sustained flooding will cause a portion of the leaves to turn yellow and drop off. The plant is taking precautions and this is normal, but again, reading the signs and correcting the conditions will keep the plant happy in the long term.

A sudden partial yellowing of leaves is normal

These curled leaves of oldhamii, one of the *Bambusa* family, on a hot day are telling us the plant is dry. This same message is sent when the plant is too wet.

during the bamboo shooting season as so much energy is being spent to produce new plants. With some species, it is also normal just before winter as the plant prepares for the cold.

Total leaf drop also occurs normally when the plant is subject to a dramatic change in conditions. For example, black bamboo is most happy in a shady environment. When a customer purchases black bamboo that has never seen more than an hour or so of partial sunshine a day, and plants it out in full sun, within days the inevitable phone call comes; "My bamboo died, I need a new one." This call is followed by the full explanation and a plea that the customer continue to water, as it will re-leaf. Many times the customer just does not believe us and stops watering, or stopped watering a week ago and is just now calling, and the bamboo dies. When he does keep watering, in a few weeks we get the more satisfying call from the amazed customer saying, "It's back! My bamboo has all new little leaves on it!" This same sudden leaf drop can occur when a plant is moved indoors or from a sunny place to a shady place. It is as if the plant grows the optimum leaf for its conditions.

The idea of a tailored leaf is not so farfetched, when you observe seedlings or young plants. The leaves may be two or three times larger in an immature plant than in the adult. One would attribute this effect to an expectation of the young one growing up in the shadow of the parent, needing all the light gathering area it can get.

Another message given by the leaves is the presence of pests or poisons. Some species like square bamboo are especially sensitive to excessive salts being deposited on the leaves. This is especially true when plants are grown with irrigation rather than rain. As dew, fog or the sprinklers wet the leaves daily, droplets form at the tip and evaporate, leaving a residue of minerals and salts. Eventually the tip turns brown. It does not significantly hurt the plant but does affect the cosmetics. We have found some species are more prone to this effect

This leaf from square bamboo shows signs of excess salts deposited on the leaves by sprinklers and fog.

than others, and that adjusting the water schedule, fertilization schedule and exposure to sun all play a role but cannot provide any cookbook formula to eliminate the problem.

The presence of aphids or mites is also seen first in the leaves. The first sign of aphids will be a black sooty mold on the leaves caused by the aphid droppings from the canopy above. Yellow tracks on the leaves are the first sign of the bamboo mite. More about both these pests can be found in the Installation and Care section.

The Culms

All bamboo exhibits a structure made up of nodes and internodes. This element is called the bamboo culm. When a culm is cut down for use, it is called a cane. The bamboo culm may have a very thin wall with each internode hollow, or may be almost solid. The spaces between nodes vary with bamboo size and type, from a few inches to a few feet. Culms come in many colors including green, black, yellow, red and a range of color combinations including

Castillon (right) and castillon inversa (above) along with several other matching species, exhibit a digital effect, where the culm color and groove color are flip-flopped. Other than this color inversion, the plants are the same - with the same size, form and habit. I enjoy mixing these matched species in plantings close to human activity, where the canes are sure to be noticed.

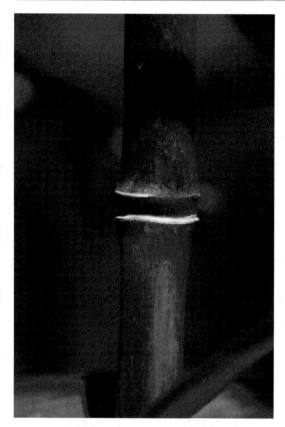

Yamadorii has green culms in the shade but in direct sunlight will exhibit both yellow and a striking red color.

Tanakae not only sports brown spots, but is known as the stockiest of bamboo. For the same height plant, tanakae tends to be larger in diameter than other bamboo.

stripes and spots. The culms are generally smooth and shiny, with branches always originating from the nodes.

Culm Aesthetics

The bamboo culm is the second fundamental element of bamboo captured so simply in Asian art. Although bamboo is not the only plant with nodes, without the nodes visible, part of the essence of the bamboo is lost. In all our landscaping projects we try to place beautiful culms within easy reach of the client, and use the bright year-round color to accent or complement much the same way flowers might be used for a seasonal accent.

It is almost impossible to walk by a bamboo plant and not touch the culm. The more it is handled, the shinier it gets and the more irresistible. A large semi-conductor manufacturer had contracted with an expert landscape designer to help him choose the best bamboo to be planted in the atrium of a new design center in Silicon Valley. After working for weeks with the designer, evaluating temperature, humidity, sunlight, habit and form, a short list was

settled on. The general manager was brought to the nursery to help with the final selection. After walking the extensive display groves he found himself in the vivax grove. He held the five-inch shiny bright green culm in his hands and said, "We

Giant black bamboo starts out green the first year and changes to black in the second as this culm is doing. In the years ahead the culm will become a shiny jet black.

Green onion (left) and crookstem (right) demonstrate interesting exceptions to the normal vertical culm.

Tibetan princess is fully red in the sun. Most normally green bamboo fade to yellow in the sun making this a startling exception.

will have this one". He had his picture taken next to the massive culm and the limo was off! Vivax was not on the short list. Oh well, he was hooked. With a few adjustments, a successful vivax planting was made with spectacular results. It was the culm that turned out to be the key factor in making the decision, not all the analysis.

There is much more to the look and feel of the culm than just the color. The node is a raised ring around the diameter of the culm. It may be the same color as the culm and barely visible or may be very distinct. The node may be marked by a ring of white powder. The node is formed as the bamboo shoot grows because it is the attachment point for the culm sheath. The sheath is a leaf-like wrapping that protects the newly created internode section of the culm. Sheaths come in a range of colors and are the most prominent attribute in plant identification. When customers bring me a branch from a bamboo in their yard and ask for identification, we can barely guess at the family, but if they would bring a culm sheath, and a basic description, we could identify it correctly most of the time. On some species, the sheath drops off within days of the shoot's appearance, and on others, the sheaths hang on for a long time, even for years. In these cases, the sheath itself becomes an important part of the aesthetics of the culm. Some show vertical veining, some are smooth and some are covered with thick rough hair. Some are all one color, white, yellow, gold, cream, green or brown, and others have bold spots.

Spectabilis as well as several other bamboos introduce a bright splash of yellow to the landscape. As this color remains brilliant all year long, bamboo in many cases can be more effective than seasonal flowers.

The spectabilis shoot (left) is very colorful. Many new shoots show a ring of white powder as seen on the mountain shoot below. Some culms keep this ring and on others it is lost.

The square bamboo shoot (left) has a longer sheath, reaching and just covering the next node. Some cover even more and others may only cover half the internode length. The vivax shoot (right) has a smooth spotted sheath.

Not only is Leopard bamboo with its patchwork of spots exotic but its shoot seems almost alien. It is an exciting day in spring when these shoots emerge, growing up to a foot a day.

Culm Sheaths

Personally, the bamboo sheath is my favorite part of this amazing plant. They are tough and flexible and many have brilliant coloring. I particularly like the sheath of temple bamboo, which has a strong purple tint on its smooth shiny inner surface. I collect the sheaths and write notes on them, do bamboo sketches on them with ink and press them. The odd sheath can usually be found on my desk or dresser. It has always felt like a crime to let them fall off the bamboo and rot in the leaf litter. I have dreams of finding some amazing use for them. It just goes to show that you can never quite predict what impact bamboo may have on your life.

For bamboo planted close to people as along a patio, hot tub or gathering area, the color and texture of the sheath can greatly enhance the excitement of shooting season. One customer ate breakfast outside on the patio next to his new bamboo planting and kept careful record of the number of new shoots and the growth rates. It was not unusual to get an excited phone call in the spring, "We have 186 new shoots so far this year and the largest is 3½ inches in diameter!" or "I could not believe it when I measured 11 inches of growth in one day. I cannot wait to get home from work to see what has happened!" It is a very exciting time, and the

The sheaths peel and fall within weeks on weaver's bamboo as with most bamboo. They decompose in the leaf litter. I like to collect them and use them for writing notes or in art projects.

sheath gives the shoot its color and texture.

For those bamboos that retain the sheath, it now becomes a key feature of the bamboo. These are called persistent sheaths. We have found some customers object strongly to a persistent sheath, while others like them. A bold almost white sheath covering half the dark green internode can make a very striking image.

Most bamboo has a smooth culm. Moso is fuzzy and soft to the touch. The fuzz is caused by tiny hairs, shown under great magnification.

Robusta's sheaths are persistent, holding on throughout the summer.

Culm Texture

Not only is the node and sheath significant, but the texture of the culm itself is important where bamboo is handled. Black bamboo becomes jet black and shiny, like a freshly-waxed car. Moso is fuzzy and soft, with a texture like velvet. Yellow groove is like the scales of a fish. It feels smooth when you rub down, but rough when you rub up. These are interesting contrasts and tactile features.

Beside the cosmetic issues involving the culms, there are other structural issues as well that make bamboo a very attractive alternative to trees and hedges.

Culm Structure

Bamboo culms are quite flexible. In high wind or loaded with snow or ice, they may bow low but do not break off easily. But more important, because they are hollow, they are lightweight. When they do break, they do not cause the damage a tree or tree limb will. When one considers the damage and decimation that follows a hurricane or tornado due to falling trees, bamboo is amazingly attractive. One customer called in a panic. A car had gone

off the road at high speed and plowed through his bamboo hedge. Could we come and replace it? No problem. Cut off the broken canes. Straighten up and tie the flattened but unbroken canes and wait. New shoots will have the hedge looking as good as new in no time! No need to replace anything. Bamboo makes a very safe barrier. The car was not harmed, unlike the result following an impact with an oak tree.

Bamboo's ability to replace damaged culms makes it very attractive in high-risk plantings, subject to vandalism or accidental impacts. If the neighbor's children cut down the 40-foot hedge, it will be back next year. If a big storm blows the plant over, just push it back upright. The long-term maintenance costs are very attractive if the right species is planted correctly.

The same flexibility that makes bamboo safe also contributes to a range of form. With a good leaf load, many times a culm will arch dramatically. This arching is attractive in some settings, where extended shade is needed or where bamboo is desired to close over a walkway or driveway. I have always wanted a long driveway lined with massive trees, completely closed over the driveway making a shady private lane. We have never lived anywhere long enough to make that possible until I modified the dream just a little. My driveway now winds through a moso grove whose giant leaning culms close over the top of the drive. It is even better than the original dream in that it is green year-round.

In some plantings, the leaning culm becomes a problem, as it may block a walkway, or hang so low as to limit access to a parked car. The best solution is to choose species that tend to remain upright for the critical vertical applications. When a culm does lean too far, you can just eliminate it or clip the top third off. The reduced weight of foliage will send the culm back up where you want it. The natural habit of the species being selected is very important when spaces are being planned. A single 50-foot tall clumping bamboo will reach out a minimum of 30 feet in diameter. Trying to trim or tie up the culms will be a constant battle. If space is tight, plant a variety that does not spread so dramatically.

Culm Fire Hazard

Little is found in the literature relating to the fire hazard associated with bamboo. When dried, bamboo canes and foliage burn rapidly. When a dried cane is burned, pressure builds up in the internode and bursts the cavity with an impressive bang. Green bamboo seems slow to burn but in forest fire conditions would burn rapidly. The big difference in comparing bamboo and trees is that the hollow canes have little fuel value. Sometimes the internodes actually have stored water in them. The leaves burn quickly and are gone. Little is left to feed the fire. This is an area that will require further evaluation before bamboo could be proposed as a lower fire hazard substitute for other types of trees.

On the other hand, the village of Saltaire, NY, with a population of 43 in the 2000 census, has made bamboo illegal within the village. Bamboo is not indigenous to Fire Island on which the village is located and the Board of Trustees has found that bamboo is "*destructive of the natural environment, indigenous flora and esthetic qualities of the village and its environs; that it*

Know your species and plan ahead. This tropical clumper dominates its space. How about this planted between the fence and the garage? Yikes!

constitutes an unnecessary fire hazard due to its highly flammable nature and in that it can inhibit the safe and timely deployment of firefighting equipment and personnel; that as such it poses a hazard to the health, safety and well-being of residents of and visitors to the village, as well as to that of both domesticated and wild animals within the village; that because of these effects it threatens the value and physical integrity of both public and private property therein; and that therefore, in order to protect and preserve said environment, lives and values, the Village hereby declares it necessary to regulate or prohibit the planting and/or growth of bamboo." Planting bamboo subjects the resident to a $250-per-day fine until the bamboo is removed. When contacted, the board stated that it had absolutely no basis for this dramatic statement about bamboo presenting a fire hazard.

Culm Harvesting

The last consideration when planning a bamboo planting is the possible use of the culms for building or craft projects. Temperate running bamboos produce very refined, beautiful canes. They are relatively thin-walled and subject to being split or crushed, so must be fabricated and joined like bamboo, not wood. Two canes cannot be bolted together without filling

the cavities or they will be crushed. They cannot be nailed or screwed together. But they can be easily split and drilled and pegged. Semi-tropical clumping bamboo has thicker walls and can be worked more like wood. The nodes are not as refined and are usually sanded off, and the natural waxy finish of temperate bamboo is generally lost. Both types have their place depending on the projects to be undertaken. In general, canes are

This attractive folding bench above is made with temperate bamboo. The canes have thin walls and are hollow. This species typically has between one and three branches per node that can be easily cleaned up. It is harder to work with than tropical clumping bamboo but is certainly more graceful and attractive.

The bench on the left is made from tropical clumping bamboo. Notice that the canes are essentially solid. Clumping bamboo generally has large numbers of branches, so the nodes are not attractive and are sanded off. It is easier to work with and less likely to split, but certainly less attractive than temperate bamboo.

cut from culms that are at least two years old for maximum strength. They should be dried fully and can be straightened by applying heat.

The Bamboo Type

To understand the bamboo types you need to understand the life cycle of bamboo. The life cycle of bamboo is one of its most tantalizing aspects and is quite different from any other plant found in the landscape. Many chapters can be written about the life cycle, but our focus will be to explain only the unique characteristics that should be understood to make effective use of bamboo in the landscape.

Flower and Seed

Like all plants, bamboo flowers and seeds, but unlike most, many species of bamboo die back

These chusquea seeds demonstrate that bamboo is indeed in the grass family. For many bamboo species, the onset of flowering and seed means the end of the plant's life. Fortunately this event does not occur very often.

Bamboo rarely flowers and seeds. Here one clump of seed is seen on a white stripe bamboo plant. A random occurrence like this may be seen once in a thousand plants and does not signify anything.

or die completely once they have flowered and produced seeds for a few years. But here is the amazing thing about bamboo. Most bamboo species flower only after many years, 20, 40 or even 100 years. Some have never flowered since records have been kept. It does not matter how often the plant is divided, transplanted or moved, its internal clock keeps track and when its time is up, it will flower and die. Although much is made of this in scientific communities, here are a few practical generalizations.

Bamboo that flowers frequently (as in once every dozen or less years) are less likely to die following flowering. Long-cycle bamboo (flowering every 50 to 100 years) may be more likely to die off when it does flower. Running bamboo is more likely to survive than clumping bamboo (see discussion of how different types of bamboo spread later). Each species or cultivar of that species is on a different cycle. This is good for the landscaper in that your risk of losing a plant due to flowering is minimal. If it is a species prone to frequent flowering, it will likely survive, and if only likely to flower every 80 years, you are unlikely to be impacted by the event! Each species is on a different cycle so mixing species reduces risk. Common species from different root cultivars may be on different cycles, so mixing the same species from different sources is also a defense.

Allergies

An evergreen plant with no pollen is really an amazing thing, a feature often overlooked among bamboo's other unique properties. Because it does not flower or set seed, no pollen is produced. For those who are sensitive to pollen, bamboo is a great option both inside and outside the house. If you are a beekeeper, however, bamboo is not the thing for you.

Bamboo spreads by putting up new shoots each year. They emerge from the ground full size and quickly grow to full height. Then appear branches and then leaves, all within a month or so. Now that plant is complete and will not grow any larger in diameter or any taller. If you top or trim the plant it will not grow back. This running bamboo, *Dulcis*, may grow a foot a day.

Spreading or Growing?

Instead of depending on annual seed production to spread, bamboo spreads underground through spreading rhizomes. Rhizomes are like a combination of a root and a bulb. They store energy and produce both more rhizomes and more plants. Bamboo spreads through rhizome branching. As the rhizome is extended, each year new plants are produced. All bamboo spreads this way. The difference between bamboo species is the distance that the rhizome spreads each year, and how far apart the new shoots will be.

But this is just the beginning of what makes bamboo unique. Normally we think of plants as growing up, but bamboo grows fastest horizontally, not vertically. It is the spreading nature of bamboo that makes it so attractive, but so hard to get a handle on. Bamboo shoots emerge from the ground full size. Yes, if a bamboo plant is 5 inches in diameter it emerged from the ground as a 5-inch-diameter shoot. The shoot grows very quickly, typically up to 6 inches a day in warm weather. We have measured up to 12 inches growth in one day. The shoot continues to grow to its full height, with no branches at all. So again, that 50-foot tall timber bamboo grew to that height in just a matter of a few weeks. Once a bamboo shoot reaches full height it begins to grow branches. It is now as tall as it will ever be during its life. Once the branches have reached full size, they also stop growing and the plant produces leaves. It is now full-grown and it never gets any taller or larger in diameter. Its growth is complete. So how fast does bamboo grow? It doesn't grow at all! But each year the new shoots are larger and cover a greater area.

The importance of this cannot be overstated. This allows a plant to be trimmed, topped or shaped just once and it will stay that way. Picture the neighborhood scene. The sun is beating down and you are hacking away at that hedge again. The top has been trimmed so many times it has become solid underneath the top layer and the huge stumps that make up the hedge can put up new branches at an astonishing speed. Eventually in disgust you have the whole hedge ripped out. Now how about a plant that once trimmed, never grows back? Not only does it not grow back, it does not even put out more branches on the top. It does tend to put on a higher density of leaves on the remaining branches, but uniformly, not just at the top. Each year when the new shoots come up, they grow to full size, but if you nip off the tip at the height you desire, they stop growing. The shoot stops right there, branches

Bamboo Life Cycle – From new shoot to finished culm in a month.   Then no more changes for years and years.

DAY
2
Shoot
grows fast

DAY
3-15
Shoot
reaches full
height

DAY
16-20
Branches
break out

DAY
1
Shoot
emerges
at final
diameter

DAY
30
Branches
leaf and
growth is
complete

The life cycle of chusquea, a South American clumping bamboo.

and puts on leaves, but at a higher density than if trimmed as a mature culm. It is finished. This can mean a major reduction in maintenance over the years, and a much longer life for the planting.

Bamboo plant size depends on the size of the shoot and the shoot size depends on the rhizome

Bamboo topped after it has branched and leaved out will remain unchanged, with the top branches making a natural looking form. No new branches will appear.

volume. The larger the rhizome volume that produces a shoot, the larger the shoot, and thus the larger the plant will be. As the years pass and the rhizome spreads, full-size shoots are produced. Typically, the smaller culms in a bamboo grove are the oldest, with the newest shoots being the largest.

If you want big plants, do not divide the bamboo plant. Each time you divide a plant, you make smaller sections of rhizome and you will get smaller shoots. So instead of one big one, you have two small ones. If instead you keep the plant together, you will get larger shoots, and the larger plant will spread more rapidly, making more larger shoots. You will fill the same area with bamboo in the same time, but the single plant will result in much larger plants, faster.

Spreading Methods

All bamboo spreads, no matter what the label says. Each species spreads in a different way depending on species type, and at a different rate based on environment. A bamboo that spreads rapidly in San Diego may not spread at all in San Francisco. On the other hand, two different species planted side by side may exhibit radically different rates of spreading. Taxonomists generally classify bamboo as clumping (pachymorph), running (leptomorph)

This green stripe vivax was topped as a shoot. Only the top two branches appeared, but the leaf density is greater because it was topped before the leaves came out. It will now look this way for life.

or as an exception that exhibits characteristics of both. These are useful classifications as they indicate the type of form and spreading properties in general, but what you really want to know is how fast and far it spreads. If that is the consideration, the two classifications are blurred even more depending on climate and conditions.

In clumping bamboo, new shoots are formed as a new rhizome grows out the side of an existing rhizome and forms a single new plant. Multiplication can occur on a dramatic scale, but the reach of each new shoot is defined by the species. Tight clumpers will form an almost solid and quite deep root ball as seen in this fountain bamboo.

Clumping Type

The bamboo type commonly called clumping bamboo produces a new rhizome from an existing rhizome with a single new culm at the end. The length of the rhizome neck determines very accurately how fast the plant spreads. For example, a very tight clumper may reach 3 inches each year. This means that a 5-gallon plant with an initial diameter of 6 inches will add 6 inches of diameter each year (3 inches per side). That cute little 5-gallon will be over 5 feet in diameter in 10 years. A clumper with a 10- or 12-inch reach will spread very dramatically over the years. The advantage and disadvantage of clumping bamboo is this predictable but inexorable growth. One does not have to be concerned about bamboo popping up all over the yard, but one must be prepared for the relentless annual increase in plant size.

Mountain bamboo is a clumping bamboo that spreads dramatically. It is just like a tight clumper but the rhizome reach can be surprising. I have measured them up to 6 feet long. Treated like a running bamboo it can be fantastic in the landscape, but planted without containment it spells trouble.

Along with this form of rhizome geometry a particular plant shape follows. In general, clumping bamboo is very tightly packed at the ground, with a very broadly spreading habit.

The exceptions to this easy classification are clumping bamboo with a very long neck. At the nursery, we planted a hillside with a range of clumping bamboo. Several species were new to us. Two years later we were quite at a loss. It seemed that a running variety must have invaded the space, as a particular clumping variety was popping up all over. We started digging these out and followed a single clumping type rhizome back to the source plants. They were indeed a clumping variety, but we measured rhizome necks over 6 feet long. This clumper (*Yushania ancepts*) turned out to be one of the most aggressive spreaders in the nursery. It is also a very beautiful bamboo provided it is contained as a running bamboo would be.

Running Type

The running varieties have a different way of spreading. The rhizome wanders great distances and branches irregularly. A running bamboo rhizome may reach as far as the bamboo is tall. In the spring, buds along the rhizome may shoot, producing new plants. The bamboo does not crowd but fills in depending on local conditions. This is its advantage and disadvantage. Great areas can be filled in quickly due to the superb spreading of running bamboo. Plants tend to remain open and spaced nicely as the bamboo shoots tend to regulate overall plant density. Its disadvantage is that containment must be exercised or unpredictable results will occur. A bamboo nursery is often asked to help with difficult removals. One such customer had an old bamboo planting that had completely occupied the area under his driveway slab.

This new spectabilis shoot has just emerged from a node on the rhizome. Rhizomes can wander great distances. Most nodes have buds and can put up a shoot or start another rhizome when the plant is ready. The plant distributes these new shoots wisely based on water, nutrients and light, keeping the grove open with a pleasant distribution of culms. Notice the roots remain short forming a dense mat on the surface of the ground, choking out weeds.

Bamboo shoots appeared all around the driveway area and popped up in all the cracks. No effort would remove or kill it. The final solution was the removal of the concrete driveway completely, along with a thick mat of rhizomes beneath it. This was a very expensive solution to what could have been an easily contained problem.

One day a lady called on the phone to ask what to do about the bamboo in her bathroom. "You have bamboo growing in your bathroom?" I asked. No, she did not want bamboo to grow in the bathroom, but "Bamboo shoots keep popping out around the toilet. I keep cutting them off but they keep coming back. I have poisoned them but to no avail. What can I do?" A little more questioning found a yard full of golden bamboo without containment, and a house built on a slab. What she did not grasp was that all that bamboo was really one big plant, and its rhizomes were running around under the house.

They found a small crack between the toilet flange and toilet and were taking advantage. You cannot stop them or poison them because they are such a small part of the whole plant that your efforts are diluted. The key was to isolate and disconnect the rhizomes from the plants in the yard, another long and expensive process.

So now you know why bamboo has a bad reputation. In its defense, the aggressiveness of bamboo varies much with species and planting conditions. Many viable containment methods exist. This is not a new challenge to the landscaper. Many grasses like Bermuda grass, ivies, creepers, blackberries and even some trees like acacias are very invasive and hard to get rid of or contain, but because these plants are familiar we are not alarmed. Getting to know your bamboo will certainly make it easy to control. If nothing else is learned from this book, knowing how to control bamboo will be a lesson well learned.

Running bamboo also has a typical form. It is generally upright with an oval shape rather than the V-shape of a clumping bamboo. It is more suitable for narrow or confined spaces.

Just as we have exceptions to the slow spreading of clumping varieties, many running bamboo in a particular climate or conditions are very slow spreading. These are a better solution where slow growth is required, for example, when we want to combine the attractive form of a running bamboo with the slow spreading of a clumping bamboo.

Clumping Compared to Running

There is great disagreement on this subject in the literature. The viewpoint very much depends on what is important to the writer. Victor Cusack in the book *Bamboo World* makes perhaps a strong case for clumping bamboo. His arguments are typical of those found in the literature. I think his passion seems to be more focused on growing bamboo to build with than using bamboo in the landscape. I disagree with his arguments when landscape is the objective, principally because I find many clumping bamboo simply unattractive to look at. In the following sections we will restate all the advantages cited for clumping bamboo provided by Mr. Cusack and either confirm them when related to the landscape or give the other side of the argument.

"Clumping bamboos grow faster when young."
This is a fact we have observed especially

in warmer climates and is one of the reasons clumping bamboo is so popular with wholesale growers. It is easier and faster, and thus cheaper to propagate a 5-gallon clumping bamboo than a 5-gallon running bamboo. That is why clumping bamboos are the first to show up at the Home Depot garden center. But we are not advocates for the grower; we care about the end customer. More important than how bushy that 5-gallon plant looks is what kind of plant it will make in your yard. A 5-gallon clumping bamboo may be a very beautiful thing, and it is, but a 10-year-old clumping bamboo is something quite different. The 50-foot tall, 30-foot diameter plant may sit on a root ball that is 8 feet in diameter and 6 feet deep, a solid mass of rhizome that requires a backhoe to remove it and is impossible to scale back. The center canes will all be dead and filled with thatch, housing untold rats and vermin. The plant could be very scruffy and unsightly. If strategically placed in a large open yard the plant would be stunning. But tightly between your pool and deck, it will be a nightmare.

"Running bamboos are invasive." Again, true enough, and the careless planting of a running bamboo should never be condoned. But do not throw out the baby with the bath water. Containment works. The question is which bamboo will best meet the landscaping need. This same criticism is the strongest argument for running bamboo. A 200-foot long privacy hedge made from clumping bamboo will require a great number of plants and cost a great deal. The hedge with the same density can be achieved with many fewer plants in much less time with a running bamboo because it will naturally fill in any open spaces within the barrier.

"Running bamboos should be contained. Most clumping bamboos require minimal clump maintenance." Yes, running bamboo should be contained, and so should clumping bamboo unless you have a very large space, but it is much harder to contain a clumping bamboo than a runner. Running bamboo tends to self-regulate, only putting up new shoots where there is room. I have running bamboo that has been in containers for almost 15 years with virtually-no maintenance and a minimum of water. The plants are well-balanced and open, with plenty of space between culms. Not only that, in some cases I have mixed two species, for example, a yellow and black bamboo. They both survived with about an equal number of yellow and

This is what that cute little clumping bamboo in the 5-gallon pot can become in just a few years. It is important to know your bamboo. What will it look like when it grows up? Generally tropical and semi-tropical clumping bamboos form a tight base with broad spreading culms. They usually need more space than a contained running bamboo.

Castillon and giant black bamboo, both running type, have lived happily together in this pot for almost 15 years with little care. Above, in 1999 about 7 years old. Below in 2007 just a little fuller.

black canes equally distributed. This would be entirely impossible with a clumping bamboo. Those we pull from the pots every 3 to 5 years, chop in quarters with an axe and Sawsall and replant. Not that I am complaining. My clumping bamboos are wonderful and I love them, but they are certainly more work to keep looking good than my running bamboo.

"Running bamboo cannot survive in an undamaged rainforest canopy." This is hardly a concern in North America. The fact that a clumping bamboo can survive is cause for concern.

"Runners are dominant, monoculture forest-forming plants." The giant timber bamboos are dominant and as a result they form great hedges and plantings excluding all weeds and unwanted additions. This makes for extremely low maintenance costs as weeds rarely find their way into our bamboo. Shorter running bamboos are shallow-rooted and make a great understory for trees, as their roots occupy a different depth. Running bamboo is more like a

Fountain is a wonderful clumping bamboo that enjoys a shady location under the redwood trees. Not much will thrive here, but because it is a clumper with an attractive open form, no containment is needed. Not much else would look this good under the redwoods.

Himalayan blue bamboo is a nice choice under a deciduous tree. The shade maintains the blue color which fades in the sun and encourages an upright form. Because this is a clumping bamboo, there is no need to cut up the tree roots installing barrier.

lawn, excluding all weeds. Clumping bamboo also excludes all from within the clump, but not the space between the clumps.

"Running bamboos are hard to eradicate." Both clumping and running bamboo are easy to remove when young, but both are hard to remove when established. Runners are shallow-rooted and distributed, so a planting of bamboo can be removed completely with hand tools. A large clumping bamboo will contain a massive root ball that will be very hard to handle without equipment. Small clumping bamboos are very easy to remove – just dig them up. (See removal tips in a later section.)

"Clumpers grow beautifully with eucalyptus or in rainforests or in a permaculture environment." Again, most North American landscapes do not relate directly to rainforests. It has been my experience that certain clumping bamboo will look very good planted under trees, better than runners of the same height. Mountain bamboo

New moso bamboo shoots are marked with the year in China so they can be harvested when they are at their maximum strength.

planted under redwood trees is stunning and just about nothing else will grow in that environment. I have some blue bamboo under a tall curly willow and it is wonderful. The shade and shelter encourage a very upright growth and the culm color shows up well in the almost total shade. This fact is especially true because of the containment issue. It is very hard to install barrier effectively among tree roots. Although short running bamboo planted for a ground cover is perfect in a forest setting, clumping timber bamboo is best if large specimens are desired.

"Running bamboos are structurally inferior." If harvesting canes is an important parameter then both the strength and attractiveness of the canes must be considered. Some of the strongest and largest canes are from clumping bamboo but unfortunately most of these bamboo are not cold hardy. When considering just what bamboos are available in the temperate range, clumpers are generally smaller and less attractive, but with a thicker wall.

"Assessing the age of running bamboo culms is difficult." It is true that the random nature of running bamboo makes it very hard to estimate the age of a culm. In China, where cane production is important, every year the new culms are painted or carved with a date code. This is an easy solution. If the concern with age is predicting when flowering may occur, the table is reversed. Clumping bamboo is more likely to die as a result of flowering than running bamboo.

"Harvesting shoots from running bamboos is more labor-intensive." This is true enough for us amateurs, but not for the professionals. I spent a day in China walking the moso groves with a local farmer collecting shoots. They are dug before they break the ground for best taste. We used a tool something like a pick or mattock. Every time the farmer scratched around he found and dug a shoot with one stroke. I dug holes all over the hillsides and found only a few without helpful hints as to where to look. The farmer tired of explaining to me how you can tell where they are hiding. Look at the existing culms, see how they are set in the ground, check the direction of the branches and bang, there was another shoot. It was really quite frustrating. Instead, if you wait until the tip emerges, it is just as easy to harvest a running bamboo, where you have plenty of room to move around. You know right where the clumping bamboo shoots are but you are working in tighter quarters.

A further minor point of difference is the fire risk. Clumping bamboo presents a higher density of combustible vegetation and thus a higher risk of fire.

So the bottom line in the anti-running argument is this; "Why bother with a running variety when you can have a clumping variety?" My answer is simple, because I find that some of the most beautiful and attractive bamboos just happen to be of the running type. I pick the best of both groups and enjoy them. Linoleum flooring may be easier to take care of than wood but I still prefer wood. Who wants second best in your landscaping?

Size and Form

The size and form of the bamboo being selected is a major selection criterion. The plant listings that follow are organized by height and form. They are grouped in two basic forms, those that are upright and suitable for a narrow hedge and those that are not. These are certainly generalizations and there are exceptions. Tall bamboo can be topped, leaning bamboo can be trimmed or tied up, but the intent of these classifications is to achieve the required results with the least effort and manipulation.

Height

The first problem with height is that the height a bamboo will ultimately reach depends on the size

In this extreme case, Tibetan princess culms start out vertical the first year and fall as the number of leaves increase the weight. The progression continues until they can even reach the ground.

bamboo just loves being in a container, and usually gives us both spring and fall shoots of amazing size (which means it is very happy).

Habit and Texture

Bamboo habit refers to the shape or form of the clump of bamboo, not so much to the individual culms. Form is caused by several factors:

Culm Density – When culms are tightly packed as with clumping bamboo, they all tend to lean away from the center of the clump, seeking light and shading the roots.

Most clumping bamboos have many branches per node as robusta (center) shows. The result is a tufting or area of high leaf density near the nodes as seen in the mountain bamboo (left) -- a very interesting effect. Most running bamboo has one to three branches per node making a uniform density overall as seen in black groove bamboo (right).

of the planting and the amount of sun. As a rule of thumb, a species will generally achieve only about half its ultimate height in a narrow hedge in full sun. Out in the open, it will branch from the ground up and will be fully foliated. In a grove, the bamboo reaches for the light and only branches and puts on leaves within the canopy.

Most bamboo will achieve its full height potential in a container or confined space. This is unusual among plants, which usually fail to reach full size in a restricted space. A few bamboos like moso, however, scale dramatically with container size. Moso requires a healthy footprint of at least 50 square feet to maintain significant culms. Even in a 25-gallon container, shoots just get smaller each year until a plant with a single culm 20 feet tall is reduced to a container with 100 culms just 5 feet tall over 5 years or so. On the other hand, square

Temple bamboo is extremely vertical with very short branches, perpendicular to the culm. It forms wonderful columns perfect for framing a structure or important view.

Culm Strength – Some culms remain straight and upright throughout their life. Others drop under the increasing load of leaves and arch toward the ground. The shoots of a beautiful clumper like Tibetan princess will be straight in the first year but will droop closer to the ground each year and may actually touch the ground after several years.

Branch Length – Some bamboo like moso may have branches that extend 4 or 5 feet out from the culm, while others like temple extend only 1 foot. With longer branches, individual culms blend together to form what could be considered a single large uniform plant. With shorter branches individual plants stand out distinctly.

Branch Angle – Some hold the branches very close to the main culm where others have branching almost perpendicular to the culm. This branch angle impacts the softness of the plant texture. The more upright branches seem to arch less at the ends and form a more structured plant profile.

Branch Number – Many temperate runners have just two or three branches per node making a

Moso is the biggest of the timber bamboo but creates a distinct layered, graceful yet visually interesting texture.. It is the gentle giant.

Henon is a giant timber bamboo but although it is refined and delicate, it lacks the visual interest of moso and commanding presence of vivax. It offers a soft refined texture.

uniform look, while others add more every year with dozens at each node forming tufts.

Leaf Attitude – Some bamboo has leaves hanging down in clumps at the end of each branch while others have more upright and variable leaves. The leaf attitude has a major impact on the texture and feel of the plant.

Together these attributes make a profile or form and create a texture. Consider the three big running timber bamboos. All are about the same size but compare the form.

Moso has straight culms that gracefully arch near the ends. The long branches also arch at the ends. The small leaves are tightly clustered on each branch which causes the branch to sag further. The result is a dramatically layered look, with dark shadows between brilliant branch layers. The arching tops provide a soft round shape and the long branches give uniformity to the grove where individual plants are less distinct than the layered effect created by the branches.

Henon has straight upright culms with much less arching and branches that are also more upright yet still long. The leaves are small but perky and

Vivax is a giant timber but is bold and commanding. Its and striking texture compliments the bold grid of this San Fra building.

Clumping bamboos generally have many branches at each node. Buddha's belly has one dominant branch and many smaller branches which lead to a somewhat disheveled look.

play. Chusquea is a South American bamboo with many short branches forming at each node. These clearly form tufts along the culm, as each group of branches are about equal in length. Now consider the *Bambusa* called Buddha's belly. It also has many branches per node but all the branches are longer, and one branch stands out, being much longer than the rest. The effect is a disheveled look and no tufting. These two similar-sized clumping bamboos make a very different statement. Where the one is graceful and organized the other is chaotic and exuberant.

The effect of clump density is also a key consideration. Ask yourself questions

upright. As a result the form is upright and well behaved but with the small leaves going in all directions, still somewhat soft but more lively than moso.

Vivax is very straight and upright much like henon but with long leaves hanging straight down in rows appearing organized, neat and sharp. It is a bold, organized and more aggressive looking bamboo, a solid and awe inspiring plant. Certainly not warm or fuzzy.

The differences between these three bamboo which are all in the same *Phylostachys* family and about the same height are like the difference in form between a willow and oak tree. Form is key not only in making sure the plant will fit neatly into the space, but in assuring that the texture works. For example, moso is superbly viewed from a distance of 100 yards – is really at its best. Henon is much weaker at 100 yards; it just fades into the background. If what you want is a green background that does not draw any attention to itself so you can show off something in the foreground, than henon is the best option. If you picked moso instead, it would steal the show and you would want to get rid of whatever is in the foreground so you could see more of the moso.

Now let's consider an example where branch number and branch length come into

This grove of golden bamboo is growing completely wild in Georgia. Although it has all but eliminated weeds, it provides a wonderful open feeling, allowing easy access while walking. This is in stark contrast to the almost impenetrable profusion of undergrowth which normally keeps the hiker on the trail.

like, "Do I want to walk around in this planting?" "Do I want the dog to be able to pass through this planting?" "If the kids' baseball rolls in here, can they get it without a chainsaw?" If your goal is to walk around in a bamboo grove, your selection must be different than if your goal is an impenetrable barrier to keep the neighbors' goats out. In general terms, the taller the running bamboo the more open space can be found between culms. The shorter the running bamboo the less space found between culms. The more cold hardy the clumping bamboo, the more space between culms. The more tropical clumpers are the tightest packed.

Do not confuse culm density with privacy. Privacy has to do with leaf density and all bamboo varieties will produce leaves where light is present, making them all good for privacy with small variations. Certain species such as black bamboo have a lower leaf density than other similar green species, providing a little less privacy and thus being best mixed with another variety in a hedge.

Fresh bamboo shoots in a grocery store in Bangkok. In Asia, fresh shoots are a seasonal treat. There is no comparison to the canned shoots we get in North America.

Edible Bamboo Shoot Harvesting

One of my favorite aspects of traveling around Asia is the opportunity to sample a great range of dishes prepared from bamboo shoots. I always insist on at least one bamboo shoot dish with every meal and sometimes find two or three. They range from sweet and crunchy to soft and fiery hot. Bamboo shoots are a great source of nutrition and easy to harvest and cook. When keeping a supply of fresh bamboo shoots on hand is a priority, all species selections must become subservient to the tastiness of the shoot and the shooting season. With a creative mix of clumping and running bamboo, fresh shoots can be available much of the year.

Such a bamboo garden design is challenging as the sequence of shooting is not consistent from place to place. Paul Whittaker in his book *Hardy Bamboo* describes a shooting sequence in the spring in the UK which is quite different than the sequence I see in my garden. Violascens is usually the first to shoot in my climate, and Robert Young is the last, many times shooting so late in the fall that the shoots are interrupted by winter and finish growing in the spring. This is because we are located in the coastal region of northern California where summer temperatures are cool.

Currently, I am documenting shooting dates and evaluating the flavor of all the species we grow as the subject of a forthcoming book exclusively about bamboo shoots.

Bamboo Case Studies

By far the most popular and useful application of bamboo is in creating privacy hedges. This represents the reason most customers seek out bamboo for the first time. We will explore this application in detail with a number of case studies but privacy is just the beginning. Almost any plant in the landscape has a bamboo equivalent. Substituting bamboo for conventional landscape favorites is great fun, but why do it? The reasons many landscape designers do not is simply because they have no experience with bamboo and have seen very few species, so do not appreciate the benefits. The benefits of substituted bamboo are many, including:

- The annual growth of new shoots allows full-size plants to be duplicated rather than having small plants just getting bigger and bigger.
- The shallow roots and ease of transplanting allow large specimens to be planted without equipment.
- Bamboo is shallow-rooted and will never send roots deep into a septic drain field. It can be planted over a drain field even in areas where building codes do not allow tree plantings but do allow grass, as bamboo is a grass.
- Bamboo is exciting and exotic.
- Bamboo offers high-density foliage year-round.
- The unique growth pattern allows for a much lower maintenance effort.

Beyond just substituting, bamboo offers unique new opportunities. Bamboo is especially effective near hot tubs, pools and water features. Following are example installations that illustrate the tremendous range and versatility of bamboo in the landscape.

Privacy Hedge

The privacy hedge is the number one application of bamboo in the landscape. It could not be better suited. Here is how a general case works.

The first thing the client does is measure the length of the fence line he wants to have the hedge planted along. Determining the linear feet of hedge is the first step in estimating cost. The length is then multiplied by the cost per foot of the installed hedge and then they have the cost of the hedge. For a "do-it-yourself" hedge, divide the hedge length by the plant spacing, typically 3 feet. Then multiply by the average plant cost, from $100 to $400 depending on size. Then add the cost per foot of barrier times the length of the hedge times two. That is the basic cost of materials.

The optimum width for a privacy hedge is 3 feet. This width provides the plants with enough growing room to fill out in the shortest possible time with the largest possible shoots. The width can be varied from just 18 inches to 5 feet with no additional cost. In extreme cases, hedges have been successfully planted in even 6 inches of space. These are extreme cases, but bamboo is amazingly versatile.

Bamboo does best in a well-drained soil or mulch. The most important thing is to keep the bamboo from being planted in standing water. If the hedge falls in a low area, either plan to build up the hedge in a berm or provide drainage. If your soil is poor, either clay or sand, supplement with a bamboo soil mix that can be added in during the installation. One bag of soil per linear foot of hedge is a good rule of thumb.

Very little care will be required once the bamboo is established. In those first years you must water often, at least several times a week in the summer. We would also recommend that bamboo fertilizer and supplement be applied at least four times a year. Gophers love bamboo and must be kept under control if you want to see your bamboo spread.

Any species can be installed in a hedge but the selection should be made based on the required height to provide privacy, available width for spreading culms, and color or texture preferences. More than one species can be mixed within the hedge. As bamboo will shoot at different times and drop leaves at different times, when mixing bamboo it is best to mix dramatically different species or species from the same family. It is disconcerting to see two medium-sized green bamboos next to each other with one just a little more yellow than the other. People do not realize they are different species and assume one is sick or unhappy.

Any number or size of plants can be put in a hedge. A very effective plan is to plant 3 feet apart, and alternate 15-gallon plants and 25-gallon plants with a 50/50 mix. 15-gallon plants are between 4 feet and 10 feet tall and 25-gallon plants are between 8 feet and 25 feet tall, depending on species. Mixing them provides immediate privacy both high and low.

Case Study 1 - New Neighbors

Dave and Judy lived in a comfortable ranch-style house built in the 60's in a quiet almost rural suburb of San Jose, California. The empty lot next-door gave them complete privacy in their backyard pool. As the years progressed, the neighborhood became quite expensive and desirable and the inevitable "infill" finally happened: A new house appeared next-door -- not just a house like their house, but a lot-filling

Above right, a new house is going in on the property line. The new wood fence does little for privacy. On the day the bamboo hedge was installed (above center) the view was broken up nicely. Four years later, the new neighbor is gone and the yard now has a new Asian theme.

Only about a foot was left to plant in after the fence was installed (left). On installation day (center), temple, henon and leopard bamboo provided instant privacy with no encroachment into the driveway. Even four years later (right) the high density hedge is still upright and well-behaved. The lower branches have been trimmed off showing the beautiful culms.

monster, with second story bays extending out and seeming to hang right over the pool.

Well, drastic measures would be required. A visit to the local planning office determined that this new building was legal and code prevented any fence over 6 feet tall. This was a dead end.

A driveway ran 18 inches from the property line, and a wood fence filled almost 6 inches of that space. It would be impossible to plant large trees there, and small ones would be slow-growing. Even if trees were planted, it would not be long before the driveway began to lift and crack.

The solution was bamboo. Because the hedge would border the driveway, a bamboo was required that would be very upright and not lean over the driveway. The size of this new house required plants over 20 feet tall.

Temple, henon and leopard bamboo was chosen and the hedge installed. Barrier was installed and the plants were trimmed to fit. The coverage was immediate and most satisfying. By the time the house was finished and occupied the windows were all but obscured. A few years later, the backyard was once again the private retreat it had always been.

Case Study 2 - The
Apartment Complex

Dennis had made several attempts to block the view of the apartment complex next door. The soil was poor and the yard very small. Various shrubs did not work out. The bamboo gave instant privacy and in just three years, the apartments were just a bad memory. Bamboo hedges

allowed this yard to be transformed. He chose vivax, temple, rubromarginata, and square bamboo.

Poor soil and little space called for some careful planning. The long apartment building (upper left) presented a formidable target to cover. Giant black, vivax, temple and arrow bamboo were all used throughout this plan. Installation day (center left) shows good coverage. In spite of the poor conditions, regular fertilizing and watering has resulted in fantastic growth. The new view from the house (lower left) shows complete coverage in very little space.   This small investment in bamboo has greatly increased the livability and value of this property.

leaves drop each year. in fact, they drop throughout the year with the heaviest fall in the the spring. The hedge is vertical, tall and green year-round - the perfect backdrop.

Case Study 3 - The Pool

In this case, the purpose of the hedge is more to define and border the space around the pool and block an unattractive view than to provide privacy. Only a narrow space was available and minimal leaf drop and shade was desired.

Bamboo worked well since only about one third of the

Case Study 4 - Hot Tub

Mary needed privacy along the property line adjacent her hot tub (above on installation day), but not at the expense of her sunny back yard. Below and right is the same hedge of henon bamboo eight years later. The neighbor is completely blocked but the hedge remains delicate, light and airy. Selecting the right bamboo and containing it properly can result in a low maintenance hedge that will last a lifetime.

Case Study 5 - Extreme Makeover

The TV program Extreme Makeover is all about instant dramatic changes on a reasonable budget. As a result, bamboo fit perfectly into the plans to create a new and exotic outdoor living space.

First the unattractive view of the neighbor's house had to be eliminated (right). It was replaced with a short hedge section of henon bamboo (below).

Shiroshima was used near the entry (center right) with its limited size and bold variegation to set the tone for the rest of the yard.

Several other problem places had to be dealt with, the worst of which was a large blank garage

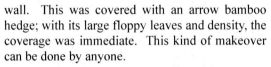

wall. This was covered with an arrow bamboo hedge; with its large floppy leaves and density, the coverage was immediate. This kind of makeover can be done by anyone.

Case Study 6 - Almost Perfect Hedge

It was the classic problem of a new house going in next door. There was plenty of yard and space so moso bamboo was chosen for its beautiful layered look when viewed from a distance. But moso is a challenge and must be handled with care. It is also quite slow in spreading.

The process was perfect. First the soil was amended (top left) in advance of the crew coming to install. On installation day the hedge was trenched and rhizome barrier was installed. Large moso plants were planted and secured to a cable running the length of the hedge (lower left). Mulch was applied to keep the ground moist.

By the second year, new shoots can be seen but they are very short. (upper right) A decorative border has been added to hide the barrier. Growth in the second year is normally minimal but more is expected in the third.

The third year results are disappointing. The moso form is starting to show with amazing density and layering, but only about 10 feet tall. Why the slow growth?

With a little understanding of moso, two mistakes can be seen. First of all, moso requires

more space to produce tall plants. The space was there, but the hedge ended up quite narrow. The barrier could have been moved out and the width doubled. That would have helped a lot. The other problem seems to be with water. Automatic sprinklers do not seem to have been installed, and from the look of the lawn, a lack of regular water has slowed the growth.

Is it all over? No, this hedge will some day reach the height of the original plants, but because of the narrow width, that may take another two or three years with adequate water.

Case Study 7 - Above-ground Hedge

Although we usually think of privacy hedges as a residential phenomenon, in this case, the managers of an attractive new building wanted to control the view and block the older rather ugly building next door.

There were many obstacles to an in-ground planting so the temple hedge was installed in 40-inch wide terra-cotta pots. The pots are touching so access is effectively eliminated. The pots are mounted on feet to assure excellent drainage, the most important factor in keeping bamboo alive in containers.

In spite of the fact that the pots cost almost as much as the plants, a hedge like this can be very cost-effective when you consider all the costs of getting permission for such an installation, and the fact that you must leave it behind when you move. This hedge can be packed up and moved to another location without leaving a trace behind!

Bamboo Specimens

Where the privacy hedge is functional, the use of bamboo specimens is dramatic. Bamboo is finding its way into commercial and residential landscape design as not only an important element but usually as the key element, or even the only element.

Bamboo offers amazing economy in large projects. Full-sized bamboo plants, 40 or 50 feet tall, can be moved into a site with only minimal equipment, and they can be positioned by hand. Trees of equivalent size may cost ten times as much and would require cranes to move into position.

Another dramatic advantage bamboo offers is its shallow predictable spreading rhizome-based root structure. We have placed bamboo on many roof structures where weight translates directly into cost. Bamboo is lightweight with very little root mass required. Its roots spread so it will interlock, providing support for the plants, but its roots do not grow as a tree's roots so cracking of containment is not a problem.

Bamboo offers the designer instant impact on the project's opening day by allowing full-sized bamboo to be planted and shown immediately, with confidence that in ten years the installation will be basically unchanged. Such a static planting would seem to be at risk from accident or vandalism, but with the annual production of new shoots, damaged culms can be removed and will be replaced each year.

Finally, bamboo is very popular with the public. People like to be around bamboo. It is as simple as that. The wide use of bamboo throughout European commercial landscapes attests to the popularity we can expect in North America. Bamboo is an attractive solution to many tough design projects.

There are problems with bamboo that should not be minimized. Most are easily overcome by education and familiarity but some are endemic.

The biggest mistakes we have seen relate to drainage and over-watering. Bamboo is not tolerant of saturated compacted earth. It requires plenty of oxygen in the soil to take up water. Another initial mistake relates to container shape. Great depth is not needed; instead more surface area is required.

In most commercial applications, rock is used as mulch, and leaf litter is not allowed to accumulate. This requires supplemental fertilization to replace the silica lost with the removal of the leaves.

Some of the harder problems are the harvesting of bamboo shoots, especially by a local Asian community who know how and when to cut them.

They are delicious to eat and rarely found fresh in local markets. In large public plantings, the harvesting of shoots can diminish the planting size over time. This problem was seen in several cases in Europe.

Graffiti is another problem invariably observed in any public bamboo planting in Asia. It is easy to scratch the surface of the culm, which creates a white scar. As those inclined to leave their mark see just one example, soon all the culms are marked! The markings do not seem to hurt the plants, but if such marking is objectionable, the first signs must be quickly eliminated so other visitors will not get the idea. Fortunately the culms are renewed each year through new shoots.

Permanent graffiti can be scratched into this smooth green vivax culm. Even scratching out the graffiti leaves a mess which can only be fixed by painting the spot green or cutting down the culm.

Case Study 8 - Perfection

"But what to do with the seven-story blank wall? This challenge ultimately became the defining landscape element of the plaza, and called for a unique and bold landscape... Bamboo provided an instant soft screen against the wall (30 to 40 feet tall upon installation) and when planted as a large grove created a unique and memorable experience for visitors who walked through it," said Christian Lemon, Senior Designer for Hart Howerton and the one responsible for this project.

Several years of effort went into producing the

plant material which was primarily vivax bamboo. Some giant black bamboo mixed in added some stunning dark canes without changing the overall look.

This high visibility design located in downtown San Francisco won the 2003 ASLA award, and was featured on the cover of *Landscape Architecture* magazine.

Shown here five years later, it is even more vibrant with a natural integration of the bamboo into a single planting. Even though this is a very windy site, maintenance costs have been low.

This has truly become a meeting place and landmark, a truly well-done landscape design.

Case Study 9 - Dual Use Bamboo

This unique medical center has a parking structure under the road. The space between the road and building is open. Tall bamboo has been planted on the ground level (immediate right) and extends up through the space looking like shorter plants in a planter from above (left and below). From the garage structure, large green canes of vivax can be seen. From above, the leafy tops show with a variety of forms and shapes.

This very unique design is built around the use of bamboo. Bamboo is not just filling a space, but is a part of the whole landscape plan. One would hardly guess that the bold bamboo shrubs shown in the lower right photo are actually the tops of 30-foot timber bamboo.

Case Study 10 - The Courtyard

Vivax was chosen for the critical planting area in the center courtyard of this upscale apartment complex. Opposing apartment porches and balconies face each other across the courtyard. Instead of seeing each other, all enjoy a vivax grove.

This is a very small space and the planter is narrow, but it can still support upright and tall specimens. Note that the full sun has bleached the canes yellow.

This planting is several years old and can be expected to increase in both density and height in the years to come.

Small Bamboo Plantings

Bamboo also lends itself to carefully manicured and sculpted designs. Small planters can be used inside or out with short bamboo to create a wonderful and easily managed statement. The the right, golden bamboo is used but cut very short.

Case Study 11 - The Shopping Center

Rather than the natural form usually selected for large commercial plantings, in this case bamboo was used to create uniform maintained spaces. Humilis, a short plain green bamboo is kept trimmed to about half its natural height making a dense tough ground cover. Aurea koi and temple have been planted close to the wall and building and trimmed so close that they look more like ivy on the wall than bamboo.

This installation clearly takes advantage of bamboo's unique growth habit, in that trimming is only needed once a year and the bamboo remains natural looking in spite of years of trimming.

Indoor Bamboo Planting

Indoor plantings can be very successful provided expectations are realistic. Giant timber bamboo can be expected to remain as installed for many years, but will not usually be able to produce large enough shoots to increase the size of the planting or replace culms that may die over time.

On the other hand, smaller shade-loving bamboo will spread and thrive indefinitely indoors.

Drainage is the number one problem with indoor plantings. In each of the plantings shown, my first question is "Where will excess water go?" Proper

Case Study 12 - Giant Atrium Bamboo

This vivax planting provides a very dramatic feel to what can only be described as a sterile and unattractive office atrium of a corporate research facility.   The bamboo effectively connects the two levels while taking up very little floor space.

Bamboo grown in the temperate coastal region of California is well suited to the temperature and humidity found inside of a commercial building.   Bamboo species that require warmer temperatures will perform poorly in the constantly cool indoor temperatures.

indoor planters should have drainage, either plumbed directly or water collection space below the plants with access to pump out excess water. Carelessness in watering could quickly eliminate the bamboo because it seems to be common knowledge that bamboo needs lots of water. All it takes is saturating a closed planting box and the bamboo will stop absorbing water. It may take weeks for the box to dry out.

Indoor plantings also require close attention to fertilization, as the plants are deprived of the naturally decomposing leaf litter and nutrients are often leached from the soil.

Installation and Care Details

Planting

Soil Preparation

While bamboo will grow in most soils it is important to remember it is not a pond plant or a desert plant. It likes water on a regular basis, particularly in summer. Bamboo likes to grow at the edge of a pond, stream or swampy area but not in it. Never plant bamboo in saturated ground. If you dig a hole and over time it fills with water, the ground is saturated and bamboo will certainly die if planted there. See the Difficult Situation section for details.

Bamboo grows in a range of soil conditions from sandy to clay. It grows best in a rich organic soil. In sandy/loamy soils, however, watering and feeding will need to be more frequent as the soil does not hold moisture as long. Supplementing the soil with a bamboo soil mix especially formulated to provide optimum conditions for maximum growth makes the well-drained sandy soil perfect for bamboo.

Soil pH is recommended between 5.5 and 6.5

This white stripe planting shows good spacing. Even though these are 1-gallon plants, because they are already full height, these are mature plants and will spread quickly. This area will be completely filled in two years.

or slightly acid for most species. Avoid using lime on bamboo. Generally, species that tolerate drier conditions may do better in higher pH soils.

Not all bamboos are salt tolerant so plant 100 yards from the ocean. Some varieties such as temple and arrow are more salt tolerant and better for oceanfront. See the detailed section on salt tolerance in the Difficult Situation section below.

Spacing

There are no rules for plant spacing. Bamboo plants can be planted tightly together for maximum density or spaced out significantly to save money. A typical spacing is as follows: Space dwarf bamboos 18 inches apart, medium and tall sizes 3 feet apart for hedges and 5 feet apart for a grove. Running bamboo tends to run uphill more than downhill so when planting a hillside, plant more plants on the lower part of the area to be filled.

Dividing

Never divide containerized plants as every time you cut a rhizome, you reduce the size shoots you will get and slow the spreading potential of that plant. Do not open up or try to spread out the root ball. Leave it just the way it comes out of the pot. It is much better to plant the whole plant as is and then divide the whole plant in a year or two when it is substantially larger. Then you will have two large plants instead of two smaller plants.

Planting

When you purchase plants they are usually ready for immediate planting unless you are instructed otherwise. If it's very hot and dry when planting, shade the plants for a couple of weeks, and make sure they do not dry out. Don't keep them waiting for more than a few hours after unwrapping and always keep them in the shade and keep the root ball moist. Do not let the fine hair roots dry out. Unwrap carefully to avoid breaking tender, new shoots.

To plant, follow these instructions:

1. Dig a hole 6 inches deeper than the container and twice as wide. If your ground is wet and any water accumulates in the hole, fill it in and plant the bamboo above ground, building up the soil around the new plant with bamboo mix. Bamboo must not be planted in saturated soil!
2. Blend bamboo mix with the excavated dirt at a one-to-one ratio for the bottom of the hole. Use straight bamboo mix around the plant. Discard the excess dirt.
3. Carefully remove the bamboo plant from the container. Be careful with new emerging bamboo shoots. These are very tender.
4. Place the plant in the hole so that the soil levels match at the top. Backfill the hole and tamp the dirt/bamboo mix tightly between the root ball and the sides of the hole. Be sure to fill all voids!
5. Build a doughnut-shaped depression around the plant and water until deeply soaked.

Pot planting is just the same except determined measures must be taken to assure good drainage. Here the pot is filled halfway with gravel before planting begins.

6. It is important to make sure the bamboo does not dry out during the first summer after planting.

Watering

The signs of drying out are apparent when the bamboo leaves roll up and become very narrow looking. A 3 to 6 inch mulch of wood chips or bark is desirable. Heavy mulch encourages rhizome growth and maintains even moisture levels.

You can check the soil to see if conditions are correct. Brush aside the mulch and dig your hand into the soil. Pick up a handful of dirt and squeeze it. If it crumbles when you open your hand, the soil is too dry. If it sticks together but your hand remains mud-free it is perfect. If it releases water or gets your hand muddy, it is too wet. A good layer of mulch will maintain these conditions.

Irrigation is necessary in the West where summers are dry, clear and hot. In the East where

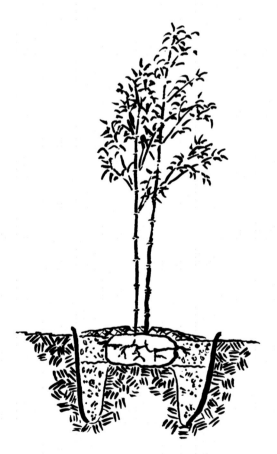

A typical planting with barrier. Roots spread horizontally so that is where the best dirt is put. Notice the cover of mulch.

summer rains are a regular occurrence irrigation is not necessary except for newly-planted bamboo or under drought conditions. Remember that bamboo is shallow-rooted. Better to water often for short durations rather than infrequently providing a long deep watering. In northern California we water about 30 minutes every other day.

Early Pruning

After planting, keep a careful watch on the plant. Clip off any branches or culms that start to dry out. In many cases the leaves on a particular culm may curl more than on other culms, or just start to get brittle. Immediately cut that culm off. If the whole plant seems stressed, cut out 1/3 of the culm and/ or cut 1/3 off the ends of each culm. It is important to get a balance between the demand for water made by the foliage and the supply of water the roots can provide. Nothing can be done to repair

This planting in China demonstrates the best way to secure bamboo against the wind after planting. The matrix of bamboo canes with cross-bracing is totally secure without dangerous guy wires or protrusions out of the planting area.

root damage during planting, and over watering will only make things worse, not better. The only recourse is to reduce the demand by reducing the foliage. Keep cutting until the remaining leaves look normal – are not curling or getting crispy. Never worry, new shoots will emerge and branches will re-leaf.

Wind Stability

Bamboo is very stable in windy conditions if the plants are well rooted or form a small grove. Over time the rhizomes intertwine making bamboo extremely difficult to push over. But when first planted, care must be taken.

The freshly-planted root ball is extremely small for the size of the plant when compared to a tree. A two-foot diameter root ball may support several 35-foot tall culms. When the ground gets wet and the wind blows, that round root ball is just like a shoulder in its socket. It simply rotates and down comes the plant. The good news is that rarely is any harm done. The plant is generally not hurt and neither is the tightly-packed root ball. All you need

to do is push the plant upright again.

It is very hard to stake a single tall bamboo plant. Its amazing height can exert tremendous loads on traditional tree stakes. Rope or wire should be attached as high on the bamboo as possible and guyed to sturdy stakes driven into the ground at least as far away from the bamboo as the height of the attachment point. These will be required through the first summer. After that, the bamboo should be securely rooted in.

If placed in a container or enclosed in barrier, the plants must be permanently supported unless the following area is provided. Bamboo needs approximately 1 foot of root ball radius for each 10 feet of height. So a 30-foot plant would need a container or contained area of at least 6 feet in diameter to remain standing in windy conditions. Plants going into a smaller area must have supplemental support planned.

Plantings which include several bamboo plants are much easier to support using traditional Chinese techniques. Plants are linked together with bamboo canes at arm's reach. We many times use duct tape to link the interconnecting canes to the bamboo plant, but they can be tied as well. Create a grid network between all the plants so that no single plant can be moved. This method is very effective and avoids the need for guy wires which extend out of the planting area.

When planting long narrow hedges in extremely windy areas, a plastic-coated steel cable can be stretched along the length of the hedge at about 7 or 8 feet from the ground. Each plant can be tied onto the cable, providing support with a minimum of unsightly rigging.

Bamboo rhizomes love to run around just under the surface of the leaf litter. A good layer of mulch encourages spreading.

Mulch

Mulch is given its own section as it is very important to bamboo health and growth. It is important not as a method of keeping weeds down, as bamboo already does that well. It is important as a method for keeping the moisture levels constant and providing an easy path for spreading rhizomes.

Bamboo loves mulch. Just leave a big pile of mulch near your bamboo and see what happens. In the spring you will find it just filled with big bamboo shoots. Mulch can be used to direct and encourage bamboo growth.

Many materials can be used. Bamboo enjoys the acidity of shredded redwood bark, and a host of other natural materials. Gravel can be used as mulch as well and even though it adds no nutrients, it provides the moisture equalization and a low resistance path for growth.

Can too much mulch be used? I have mulched several feet deep with great results. Go for it.

The bamboo leaves also form excellent mulch. In addition the leaves provide a very important source of amorphous silica, something not generally available to the plant. If the leaves are not allowed to compost under the bamboo, supplemental fertilizer with amorphous silicate must be used.

Size and Source of Plants

Field-dug Versus Container-grown Plants

Not all container bamboo plants are equal. The growth rate of a field-dug plant is much faster than the same size plant which was nursery propagated. Nursery stock is immature bamboo shoots from immature rhizome. It may take many years until you see mature shoots. Field-dug plants are small shoots from mature rhizome. As the rhizome spreads out, full-size shoots are possible much sooner. Do not be fooled by lower-priced mail order stock! Check the source and always get field-dug plants if possible.

Size Economy Is Unexpected

Even before a supplier of bamboo has been selected, a plan must be developed and the plant sizes selected. Let's say we have 100 square feet to fill with bamboo at least 15 feet tall around our new hot tub. Do we want ten 5-gallon plants (around 4 -6 feet tall) or two 25-gallon plants (around 15 feet tall)? They may cost about the same but they will certainly not look the same when planted, or in the years to come.

The economics of large vs. small plants are very much leaning toward large plants. This goes back to the discussion earlier about how bamboo spreads. Small plants will never grow larger and they produce small shoots in the next years. Large plants already look good and will produce larger shoots in the following years. In three years, in our example planting, the two 25-gallon plants have filled the entire space with 10- to 20-foot bamboo plants. The same space planted with 5-gallon plants will also be full in three years, but with plants from 5 to 10 feet tall. Both sizes spread rapidly and both will produce larger shoots each year, but the rate of spread is much faster than the rate of height increase. Whatever the case, you are usually ahead with the biggest plants available.

Another factor relating to plant size is the risk factor. A small plant can be dinner for a gopher in just one day. Gophers can slow the spreading of large plants but not kill them. Large plants rapidly spread, even in marginal conditions. Small plants die if conditions are harsh.

For the same dollars spent, privacy is created with large plants in about 1/3 the time as small plants.

Containment Strategies

Like all landscape design strategies, you can try to force plants to be the way you want them to be or you can take advantage of a plant's natural habit to achieve the desired goal. So it is with containment. The best strategies are passive, through selection of the right species and proper placement and conditions. In some cases artificial means must be undertaken to assure the best end result.

Bamboo rhizomes generally travel in the top 6 to 10 inches of the soil. When they hit a barrier which deflects them upward they break out of the ground and form a new shoot. If deflected downward, they can burrow quite deeply. A running bamboo can generally project a rhizome the distance along the ground equal to the height of the bamboo. Rhizomes tend to run uphill faster than downhill. Bamboo generally cannot spread or even live in standing water. With these facts about

how bamboo makes its living, we now can cover a range of containment strategies.

Natural Barriers

The best natural barrier is the environment. Water makes a great natural barrier. Creeks and springs can be used to create natural segmentation of your planting area. Bamboo must be planted well above such wet areas so some berming or trenching may be needed.

Restricted watering can control bamboo in locations with a dry summer climate. If the climate is dry in the summer, an effective barrier can be accomplished through selective watering only near the center of the plant.

Moderately spreading bamboo planted in a dry area with natural rhizome trimmers like gophers will remain in a clump for many years without any containment or spreading. Plant the same bamboo next to an irrigated and fertilized garden, and it will quickly take over the garden. Control of water is a great way to direct bamboo.

Clumpers can prove the hardest to contain. It is best to plant them where they have adequate room to spread, or switch to a runner.

Another natural barrier is to harvest a bamboo that spreads at a rate that matches the new growth. As canes die or are cut for building projects, or as shoots are harvested for eating, new shoots

Bamboo rhizomes love to run around just under the surface of the leaf litter. A good layer of mulch encourages spreading.

will replace them, resulting in a balance. Select a decorative bamboo that is not particularly aggressive and a little maintenance will keep it under control. A shallow trench one shovel-width wide and 8 inches deep will block 90% of bamboo expansion. Clumping bamboos are especially easy to control by always harvesting the shoots around the perimeter of the plant.

An unexpected way to contain running bamboo is to sterilize it. This is done by removing the rhizome from the plant before planting. With the rhizome removed, the plant will live on but no new shoots will ever come up, and no new rhizome will be formed. This method is especially prized in Japanese gardens where change and growth is not appreciated. Once the bamboo plant is trimmed perfectly and planted, it never changes, year after year. Of course the plant will die in ten or fifteen years without the new shoots to continually renew it, and at that point it would be replaced with a substitute, as close in size and form as can be found.

Artificial Barriers

Through many years of experimenting, many materials have been proposed as a barrier but the only material we have found 100% effective is 60 mil thick polyethylene sheet, in rolls at least 30 inches wide. It is UV protected for long life and has a plasticizer added to make it easier to work with. Concrete cracks, metal rusts, most plastics age and crack, rubber is too soft, and rhizomes are sharp and can penetrate a surprising range of materials. We have had customers bring us polyethylene barrier that is .040 inch thick with a rhizome right through it.

Barrier is installed by trenching a perimeter 27 inches deep around the planting area and then placing the barrier on edge. Tip the barrier away from the plant at about

STEP 2
Cut about a 15-degree angle on the trench, sloping outward. Notice below that when the soil changes and becomes hard it is reasonable to penetrate the hard soil, but not dig full depth.

STEP 1
Dig a trench 27 inches deep around the perimeter of the area you want to contain. It is easier to remove all the dirt when the area is narrow.

STEP 3
Lay the barrier in and press tight with your foot to assure that it is against the outside wall and backfill. Let a few inches of barrier extend above the surface and encourage the bottom edge to lie flat in the bottom of the ditch.

STEP 4
To keep your slope in the corners, put a crease in the lower part of the barrier and allow it to overlap.

STEP 5
Finish the installation by overlapping and mechanically clamping the ends of the barrier. Two redwood 2x4s on opposite sides screwed together works well.

a 15-degree angle. Let the barrier stick 3 inches out of the ground. Support with a decorative wall or border if possible. Try to have only one seam. The seam should be overlapped and mechanically clamped.

Planting in Containers above Ground

Another solution is to plant in a container. Use a container large enough to have a space at least two inches between the edge of the root ball and the side of the container. Squatty tub-like containers are better than tall deep ones. You may plant in a half-barrel (wood or plastic). All containers must have large drain holes. Fill the bottom half with gravel to assure good drainage. You may sink plastic or wood containers into the ground with about 4 inches of barrel above the soil line provided you have made provision for drainage. Bamboo soil mix should be used to fill the container. Good drainage and moisture retention is essential.

The potted hedge fits a very specific set of needs. It is of great benefit when:

- You do not have access to the soil such as on a concrete pad, around a pool or walkway.
- Your soil is very saturated and drainage would be a problem.
- You need to be able to move the plants. For example, if you plan major construction in the next few years, the hedge can be moved out of the way and placed back as needed for privacy.
- You rent and cannot plant bamboo, or plan to move soon and want to take your hedge with you.

Bamboo in pots makes an immediate and complete above-ground hedge. In this case, it was a temporary hedge used for a special event.

- You need privacy for a special event.

Maintenance and Fertilizing

Although any grass fertilizer can be used with bamboo, bamboo fertilizer and supplement has all the micro ingredients needed specifically by bamboo for long-term health. In Japan, tests revealed three times the growth on regularly fertilized groves compared to unfertilized ones. Fertilizing the bamboo two or three times per year will keep it healthy but for dramatic growth, use bamboo fertilizer and supplement every time you water! Of course bamboo just loves composted manures of any kind, just don't overdo it! Wet, sticky manure that is not fully composted will compact and seal the ground. The ground must breathe or the roots will rot. Our supplement will also make your bamboo more resistant to aphids.

Do not worry much about over-fertilizing. We have tried grossly over-fertilizing with ammonium nitrate to see what would happen and the bamboo showed no stress. It seems to have a high tolerance for chemical abuse because of the distributed nature of the rhizomes, which is also why it resists poisons like "Roundup" so well.

Thinning, Pruning, Shaping

The key to keeping the bamboo beautiful is in how one prunes and grooms. Remember that once cut, culms will not grow back. Branches may be trimmed off to expose the culm. They also will not grow back. Green culms in full sun will yellow over time. There is no way to prevent this and it has been observed on all green species. For a more open look, the older culms need to be thinned out and the spaces between them kept even and open.

For the largest culm diameters, cut all small new shoots each year. It is best if you simply break off the shoot as soon as you see it emerge from the ground if the diameter is not large enough. In that way energy is not being wasted growing the shoot just to have you cut it off when it is finished. Only let the largest shoots grow to full size. Go through the grove and cut away all small culms.

When cutting a cane, cut just above a node so as not to make a hollow pocket that will collect water. Let the cane lie in the grove until the leaves fall off so that they will return nutrients to the soil.

Pests

Bamboo is amazingly free from pests in North America. Most people never find the need to treat the bamboo for bugs or limit access because of grazers and nibblers. Three pests are occasionally seen. The most common and irritating is the gopher. Aphids and the bamboo mite are the other two.

Gophers are a common pest on the West coast. They love bamboo and will keep your plants from spreading by eating the new shoots and trimming the new rhizomes. In that respect they are not a pest at all unless you want to see the bamboo spreading, then they must be controlled. We hear an amazing range of stories from customers about the battle they have with gophers. Some "tricks" may be fact but some are clearly fiction. The good news is that gophers are somewhat territorial and therefore do not just keep multiplying to the complete destruction of your plants. Once a bamboo

Where harmless moles leave raised tracks, fresh dirt piles are a clear indication that gophers are hard at work below ground. Gophers love bamboo and can eat a 5-gallon plant in less than a minute.

planting is established, the presence of a gopher in the neighborhood has little impact. Where they can be dangerous is in a new planting of small plants. I have seen a 5-gallon plant pulled down a gopher hole in less than one minute. So the real problem is with new plantings, and the solution is to wrap the root ball with gopher wire – a chicken-wire type material with smaller holes. Once the plant is established and putting new rhizomes out it should be able to produce enough plant material to keep the gopher happy and still expand. If that is not adequate, many other solutions can be sought, from traps to noise makers and poisons. The only success we have had in a nursery situation is feeding them vast quantities of poisoned bait, a time-consuming and distasteful solution.

Aphids are a minor problem found occasionally with bamboo. The aphids secrete a sweet material that promotes the growth of sooty mold, resulting in some black deposits on the leaves and anything under the bamboo, like your car. The aphids will not kill the bamboo but do look unsightly. The first sign of aphids is the appearance of the black deposits on the leaves as the aphids themselves are usually up higher, are small and hard to see.

One's philosophy concerning pests dictates your course of action. We believe that bug-free plants are unnatural and that control should be the goal, not eradication. For complete eradication, various spraying methods can be employed. It is very hard to coat the leaves on such tall plants, so the best methods apply a liquid to the roots. It is taken up by the plant and poisons the bugs. Other methods that use soaps and oils are widely discussed in the literature but cannot be easily managed on the large scale of a nursery so I have only limited experience with them.

We think a better plan is to control aphids by releasing ladybugs and predator mites each year. Although not eliminating the aphids, the population is controlled. Another defense is a strong offense. Healthy plants are more resistive so be sure proper nutrients are provided to your bamboo. This does not eliminate the aphids but reduces their numbers to what by some is considered a reasonable level.

The other pest now showing up is the bamboo mite. This is a more dangerous pest. The mite can be identified by a row of small rectangular pale spots making a track down the bamboo leaf. The mite makes a webbed pocket on the underside of the leaf where it lives by sucking sap from the leaf. The webbing makes this bug even harder to kill with contact poisons. For a limited outbreak, the best strategy seems to be to cut off the infected canes and burn them. The bamboo grows back and the pest is eliminated.

If you do have existing bamboo and are purchasing new bamboo or getting some from a friend, be sure to look it over for pests before introducing it to your yard.

Removal

Bamboo has a reputation for being very hard to remove. Large clumping varieties must be treated like a large tree, with the canes cut down with a chain saw and the massive root ball ground up. Unrestrained running bamboo is generally dug out and the ground cultivated to loosen all the rhizomes. They can be raked from the ground and eliminated. It is hard work but can be accomplished with hand tools. Poisons are generally reported to be ineffective.

The best way we know to kill bamboo is by using your knowledge of its lifestyle. Bamboo is very susceptible to drowning, and can be easily eliminated by prolonged flooding. We found this out one winter when a drain plugged and water backed up into a bamboo grove. In a week the curled leaves were sounding the alarm. In two weeks the leaves were yellow and in three the entire section was dead. We cleared the blockage but the bamboo never re-leafed or came back. It was entirely dead. So, during the time of year when you are likely to have rain anyway, berm the area around the bamboo and flood for about two or three weeks. It must stay underwater the whole time, so use a hose as needed. The results are reliable and total. The entire plant and rhizome network is permanently killed with very little effort.

Special Planting Conditions

Dry

The common running timber bamboo is native to a coastal region of China with about the same rainfall as northern California, but the rain falls almost entirely in the summer instead of in the winter as it does in California. These bamboos are quite drought tolerant in the winter as they are superbly adapted to that native climate. When grown in a place with a long dry summer they are more susceptible to drought than some native species. I have, however, seen stands of bamboo growing in northern California with absolutely no irrigation. Although the bamboo clump is not spreading, it looks great. But for a new planting, water must be provided. The most significant step that can be taken to reduce sensitivity to drought is the use of a heavy mulch, an effective solution regardless of the species.

When seeking a plant with drought tolerant requirements, consider a bamboo that originates in a climate similar to your planting. Bamboos are available from all over the world, and from climates of every conceivable type, certainly from every climate found in North America. There will be a species that meets all the requirements of any

indigenous species. All that may be needed is a little education at the local planning department.

For example, I have seen chusquea bamboo that originates in South America growing well along fence rows of the central plain in the hot dry summer in Chile where the climate closely matches that of central California. It is a good choice when watering may be sporadic.

Arrow bamboo is thought to originate in the islands of northern Japan. These islands receive about half the winter rainfall during the summer months. As a result, this bamboo is also very resistant to summer drought.

Wind and Salt

Bamboo is certainly a sturdy plant and well suited for planting in windy locations. Certain species, however, are particularly durable, and able to handle the extremes of wind and salt found near the ocean, both salt in the ground and airborne salt carried by the wind. These species of bamboo have adapted to rugged island and coastal environments which include extreme wind and salt such as those originating on the Pacific coast of South America and the coasts of China, Taiwan and Japan. These species generally have a thicker leather-like leaf and perform well in beach front plantings.

The chusquea bamboo is again a star in this regard. I have seen it growing on the beach in the sand just above the high tide mark in Chile, subjected to the full rage of the Pacific Ocean.

Members of the *Pleioblastus* family such as *hindsii*, *simonii* and *linearis*, as well as arrow, are thought to be native to the Japanese islands and are also very salt tolerant as well as very effective windbreaks. We have used temple and yamadorii in beach front plantings with success. Some say yellow groove and spectabilis are good as well but we have not confirmed this.

Arrow is a great bamboo for tough conditions. It is draught, wind and salt resistant.

A number of the bamboos from the *Pleioblastus* family like *hindsii* above and the *Semiarundinaria* family are excellent in windy and salty conditions.

Wet

Bamboo loves water and responds dramatically when watered adequately but it does not like having its feet wet. Bamboo depends on air in the soil to produce the reaction that allows water to be extracted from the soil. In saturated soils bamboo will die. When planting in saturated soil, or soil that is seasonally saturated, precautions must be taken.

We did one very attractive planting around a swimming pool in Hollywood. Within a month we were called back for a warranty replacement. This was a surprise and concern. The plants were indeed dead and no possible explanation could be found, and new plants were installed. As these started to die, we became really concerned. When we visited again, we found the soil quite wet. When we took a shovel full of soil out and waited a few minutes, the hole began to fill with water. Ah, we found the problem. The ground around the pool and patio area was captured with retaining walls, patios and planters. When irrigated excessively, there was no place for the water to go and the ground became saturated. During our first replacement the irrigation had been turned off as the plants were all dead, so we saw no problem. Saturated soil is the worst condition for growing bamboo. We had to remove the bamboo and install drainage pipe and gravel below grade, and then slightly berm the soil in the planting area. That took care of the problem and the hedge was healthy and happy. So in wet conditions, the best solution is drainage and berming, so water cannot accumulate.

Nature has also provided at least one other solution. Water bamboo (*Ph. hetroclada 'Purpurata'*) is unique in that the rhizomes are hollow and carry air to the roots allowing it to survive where other bamboo cannot. On the nursery we had a low spot in which nothing ever looked good. The ground was compacted and was not sandy like most of the nursery, with little mulch. The bamboo would inevitably look poor and rhizomes tended to run on the surface. The ground was always quite wet. We tried water bamboo in this spot and it is very happy.

Other bamboos that we have seen somewhat tolerant to excess moisture are square, palmata, veitchii and arrow.

Water bamboo has a hollow rhizome allowing air to reach the roots. As a result it can tolerate much more saturated ground than most bamboos.

Cold

All the species lists, books and web sites list minimum temperatures in which a species can be expected to survive. It has been our experience that this is almost as useless as trying to pick a minimum temperature at which people will survive. It really all depends on how you are dressed, how long you are exposed, if you are wet, and if the wind is blowing. It is the same for bamboo, where instead of clothes, protection from buildings, other trees or mulch are the key factors.

For plants which cannot survive frost, this effect is not as significant, but for those that do tolerate frost, the actual amount of frost seems to vary both with conditions and the particular cultivar. We have purchased the same species of bamboo from different wholesalers and had the entire load from one source completely die the first winter, and the plants from the other survive our winter. All nurseries know exactly what will and will not survive in their climate but unfortunately have little precise knowledge of what will survive in a colder climate.

The winter of 2006 saw all my blue bamboo dead except a few plants under the cover of a large tree. Having recorded the temperature, we then modified the minimum temperature on our web site. But we have had colder winters without any damage to the blue bamboo. Even with firsthand observations of lows and highs, other factors than just the low temperature are most likely responsible. That winter, although not producing new lows for our area, saw an unusual stretch of freezing weather lasting for over a week. Our koi pond froze over to the extent the kids walked around on it. I do not ever remember seeing it with more than a thin ice during a cold night and gone the next day. So our blue bamboo could not handle the duration.

The same kind of factors relating to wind and humidity may also dramatically influence cold hardiness.

So the lesson to be learned about cold hardiness is to buy your plants from a nursery with similar climate to your own, or buy species you observe looking good in the area. Buying bamboo from Florida for planting in Ohio is much more risky than buying locally even if the list says it is OK.

Tibetan princess looks very different in a sunny location (above). On the left, with only early morning sun, the branches are heavy with leaves and arch beautifully.

Shade or Sun

Just as in cold tolerance, all the lists and books specify a light ranking, either on a scale of 1 to 5 or a shade, partial shade and full sun kind of scale. This measure is another that we have found quite meaningless. Perhaps a bit of an overstatement, but in our experience, in our relatively cool coastal climate, all bamboo will grow tolerably and look good in the shade. Period. I have also found that all bamboo will grow in full sun, but very clearly, some bamboo does not look as good in full sun as it would with some shade. Having said that, there are of course extreme cases.

Square bamboo is superb in even dark conditions. While visiting a temple in China I was moving through a very dark corridor. A small chink in the roof let in just enough light to feel my way along. I ran into a bamboo plant growing inside. I could not see it well but felt it and noticed the distinctive square culm and protrusions at the nodes. Whether in the ground or a pot I do not remember, but I was impressed it was alive at all.

Now I have square bamboo growing in three very different conditions at the nursery. One square grove is growing in a break in a live oak forest at the bottom of a valley. It gets very little sun and looks great. Part of it is in full shade. The leaves are a rich dark green and full. Another grove is on flat land in full sun. It has grown tall and tight and also looks great, contrary to what you would expect. A third planting is on the south facing a wall of the warehouse in full sun. It is also alive and growing but leaves on the crown of the front row of plants are bleached almost yellow and it is a very unhappy looking clump of bamboo. The sun beats down on the ground under the plants as it is a relatively new planting and not making its own shade yet. So even square bamboo which is widely known to be an excellent bamboo for shady conditions can look good in full sun if allowed to shade its roots, as in my second case. But in an extreme case with full southern exposure and the additional heat of a dark parking lot and dark warehouse, it is unhappy.

Some other bamboo looks much better in shade than full sun. The rich green and white stripes of the variegated white stripe bamboo fade to a dull yellow and cream striping after a full summer in the

Square bamboo in full sun with a little protection and shaded roots (above) and with full exposure and no protection (right). It is still alive and doing its job, but it is not very happy.

sun. So also with other variegated short bamboo.

Several bamboo look so different in sunny and shady locations that it is hard to believe it is the same bamboo. Two examples come to mind.

Tibetan princess is grown along the north side of the koi pond. Directly behind it is an extensive towering moso grove which provides full shade. Each culm is green and burdened with masses of leaves to the extent that the culms arch down to the water. It is so dark green that it almost looks blue green. The sight of this bamboo in the early morning light with heavy dew on the leaves and the pale backdrop of the moso is one of the most beautiful sights I have seen. The plant is extremely happy producing an abundance of new shoots each year and even though a clumper, it spreads at an alarming rate. Clearly this plant was suited to this environment.

I so loved this plant that I removed half of it and split it into three plants which I placed in a peninsula protruding into the picnic area where it could become a major focal point. This location was in full sun with a southern exposure. A year later I did not recognize my beloved plant. This transformed plant had brilliant red canes, a much lower density of green leaves with culms that hardly arched even half way to the ground. Hundreds of new shoots emerged and in the second year each plant was bigger than the original. Obviously the plant was very happy in this sunny location and became another favorite of mine all over again.

Another unusual case we found was with yamadorii. One 25-gallon plant of this quite rare relative of temple bamboo was found and identified. It was planted in a cool shady location near the temple grove. Knowing very little about it, we watched to see what it would do. Graceful green culms with long leaves emerged at a startling rate. Soon the space was filled with this luscious bamboo which was not behaving at all like the very slowly-spreading temple (in such a cool location). So much was produced that we moved some up the hill to a warmer microclimate and planted it in full sun. There it also spread with almost alarming speed but besides the green culms, anywhere the culms were in the sun, they produced both yellow and purple reddish-colored culms. No pattern could be discerned as to why one culm might fade to a yellow and another to a reddish color. What an unexpected result. So should this plant be ranked

White stripe bamboo is a great option for shady conditions. It also does well in a pot, indoors or out. This pot is in full sun and the leaves are bleached but still happy. In shade (inset) the leaf color is rich.

for shade or sun? With most bamboos, the only answer is yes.

Indoor and Container Planting

Bamboo responds very well to container planting. Even large timber bamboo can be maintained for many years in quite small pots. There seem to be a few tricks to keep the plants looking good.

In general, clumping timber bamboo will outgrow the pot more quickly than running bamboo. Running bamboo can be kept in a pot a a for a very long time by not overwatering and fertilizing it once it is full-sized and the pot is full.

Bamboo species can also be mixed in the pot without harm.

one pot a little over 3 feet in diameter that has had black bamboo and castillon bamboo (with a bright yellow cane) growing together for almost 15 years. During that time nothing has been done to it other than the trimming of dead canes every few years. It is hardly watered and rarely fertilized and thus puts up 2 or 3 new shoots each year, just enough to keep it looking fresh. It looks the same today as the first time I saw it. Perfect.

Some bamboo for some reason does not do well in containers. Moso is a bamboo that does very poorly. Large plants cannot be maintained in containers. Some bamboo like square does better in containers than in the ground, it seems. I have seen a 15-gallon pot of square produce 30 new shoots in the spring, and some more in the fall.

The results for indoor planting are just as mixed as every other aspect of bamboo. Some are known to work well indoors, but we have had good success with others that would never be considered good candidates.

Bambusa multiplex, borinda, Mexican weeping bamboo, albostriata and veitchii are all considered good small bamboo for indoor planting, but most customers want to see big canes. Mountain and arrow are larger bamboos known as good options for indoor planting. We have planted many large running bamboo indoors such as black and vivax with good results. But this is a good time to give a word of warning. Large bamboo indoors in most cases is not to be viewed as a sustainable planting. A number of years ago the San Francisco Airport placed very tall timber bamboo in planters in an atrium area. The four-inch canes, grown in the humid tropical conditions of Florida, had massive height, soaring several stories in the front of the international terminal. But taking these plants from Florida and placing them indoors in San Francisco

is like putting them on ice. They have held their leaves but each year a few canes have died, and the new shoots produced were less than an inch in diameter and only 8 or 10 feet tall. Each year the clumps have been getting smaller. Although a species that thrives on the cooler coastal climate of California would have held up longer, it must be conceded that any indoor timber planting is a disposable planting, and replacements will be needed from time to time. Just as most indoor potted plants are rotated out with fresh plants, so the timber bamboo would need to be rotated out. To keep consistency, such a requirement should become part of any major purchase contract.

One cannot expect these plants to continue to put up 40-foot shoots, but they can be expected to remain as they are for a number of years, with the occasional loss of a culm. At some point they will need to be replaced.

This 5-gallon albostriata makes a great houseplant. The biggest problem with bamboo indoors is too much water. Use good bamboo potting soil and plenty of drainage.

Species Selection for Harvesting

Planting an area of bamboo as a source of canes or shoots to eat is an attractive investment. Canes and bamboo shoots are quite expensive and supply is very haphazard, as wholesalers rarely can keep an adequate stock. Harvesting from your own grove keeps your bamboo healthy and attractive with very little effort. Selecting bamboo to suit your palate or hobby will not only provide excellent landscaping options but provide years of savings, and the satisfaction of harvesting your own bamboo.

Building Projects

Bamboo is a challenging but rewarding material to utilize in buildings. It is best used as a structural element where its amazing strength and light weight can be capitalized on. It is hardest to use as a decorative element because it is round,

A few canes harvested from your own bamboo clump can be turned into art, even without carpentry skills.

hollow and tapered. That makes it hard to work, fit and attach. Unfortunately there has been no standardized grading of bamboo and it is not recognized by building codes as a structural material. It is used extensively in South America

Working with bamboo is different than wood. Here a standard joining method uses a bamboo pin to lock the smaller cane into the larger cane.

for structural elements, most impressively for giant overhanging roof structures, but attempts to use it in Europe and America have been unreasonably expensive as cities and counties have been overly cautious about issuing permits.

So bamboo is a great material for small non-permitted structures like fences, gazebos and garden sheds. Bamboo is not used like wood. It is used more like metal tubing which cannot be nailed together but must be fastened with care.

Canes should be selected which are at least two years old, dried and straightened. The application of dry heat to the bamboo allows it to be bent quite

Here is a great way to make a boring redwood fence into something fantastic. There is always a way to make use of bamboo once you have it growing nearby.

easily. Most running bamboo can be easily split as well as some of the thinner wall clumpers.

Bamboo is as susceptible to termites as pine and should be kept out of contact with the soil.

Many resources are available to help the beginner learn the way bamboo is used in building.

Crafts

Bamboo is used widely in art and craft projects. Clients come to the nursery to select specific bamboo for use in bicycle frames, earrings, flutes, fishing poles, walking sticks, archery bows, wind chimes, basket making, kitchen items and martial arts accessories. The range is overwhelming, and with a little creativity you can probably find a way to apply bamboo to your or your client's hobby.

Again, many books and resources are available to aid in craft making, but for those who have an interest, the key is planning ahead and getting the right bamboo in the ground as part of the landscape.

This entry was crafted with local bamboo in Japan.

Make your own instruments. How about a flute or *didgeridoo*?

country and insist that only bamboo shoots from this particular species are worth eating. This bias is most evident when comparing cuisine from the tropical regions of southern China where clumping bamboo is preferred, and the cuisine from the temperate northern China and Japan where running bamboo is used.

Now add shoots from non-Asian species and the mix is complex. I think they are all great-tasting, although different and fun to eat.

Black and leopard bamboo are used to make this high-performance mountain bike.

What a treat to see two different kinds of bamboo shoots laid out with the fresh vegetables that would be making up our lunch in this country restaurant near Dragon Mountain in China. Lets eat!

Shoots

Almost all bamboo shoots can be eaten. Some, however, taste better than others. Some can be eaten raw directly from the ground, where others are bitter and need to be boiled and the water discarded before eating. For those with a taste for bamboo, selecting species known for being good to eat is a wise idea, but seasonal considerations should also be made. Shoots are best eaten fresh, so species should be planned to provide fresh shoots as much of the year as possible. Even the same species can be positioned in warm or cool spots to shift the harvest time.

The early shooters can be placed in the warmest location to start the season and the late shooters in a cool place to extend it. It is thought that soil temperature triggers shooting time, which is a complex function of daytime and nighttime temperature ranges.

A garden optimized for shoot production will contain both running and clumping bamboo. Most bamboo cultures are very narrow-minded about what constitutes a good shoot. They are very partial to the shoot from their garden in the home

Our Favorite 50 Bamboos - The Species List Organized by Size and Use.

SMALL BAMBOO 1'- 4' TALL

SMALL BAMBOO 6' - 10' TALL

MEDIUM HEDGE BAMBOO 12' - 25' TALL

MEDIUM SPECIMEN BAMBOO 12' - 25' TALL

TALL HEDGE BAMBOO 25' - 40' TALL

TALL SPECIMEN BAMBOO 25' - 40' TALL

GIANT HEDGE BAMBOO 40' - 60' TALL

GIANT SPECIMEN BAMBOO 40' - 60' TALL

White Stripe - my common name (used throughout this book)

(Dwarf White Stripe) - widely recognized common name (same as my common name when available)

Pleioblastus fortunei variegatus - Latin name (usually shortened to form my name when no common name is known)

Small bamboo under 4 feet tall are generally of the running type and are very useful as a ground cover, either alone or under a canopy of trees or tall bamboo. They usually look better with a little shade. They also can be cut to the ground each winter without causing any harm. Because the rhizomes are small and shallow, they are quite easy to contain.

The bamboo in this section between 6 and 10 feet tall contain both clumping and running bamboo. They are most often used like any ornamental shrub and can be trimmed and formed or allowed to grow naturally. Because they are bamboo, they generally stay looking better with less effort than traditional shrubs.

The medium-sized bamboos are the real workhorses providing the perfect-sized plant on the day installed and will remain perfect-sized forever. This is a major advantage of bamboo.

The medium hedge bamboo are all runners, optimum for filling a contained space quickly to provide interest and privacy as quickly and cost effectively as possible.

The medium specimens are mostly mountain clumpers which combine cold hardiness, grace, and a tight, slow spreading habit. They make impressive focal points in any landscape.

The tall bamboo must be thought of more as trees, than as a grass. But unlike trees, they can be cost-effectively purchased at full height and easily handled and installed. The tall hedge bamboo are upright and moderately spreading, so will fill in a hedge quickly, even when space is very limited. Tall hedge bamboo can be considered unique in its ability to provide such an impressive year-round shade with so little maintenance, in such a narrow restricted area.

Tall specimens include both running and clumping bamboo that by their very nature do not provide either the narrow uprightness or density of foliage to be suitable for hedges, but are still very attractive elsewhere in the landscape.

The two giant hedge bamboos are truly remarkable and make up a large part of the privacy hedges installed. Even when the height may not be needed, the large canes are very desirable. The other giants include a running and clumping variety. They dominate their space and present a very attractive alternative to trees, especially in the speed with which they become full-sized.

This clump of Robert Young is so slow running that the canes form a living privacy wall. The classic designation clumping and running bamboo does not really provide a picture of what the bamboo will do in your situation.

Our Favorite 50 Bamboos

Of the thousands of species of bamboo, about 400 are listed by the American Bamboo Society as available commercially in the U.S. This is far more species than a landscaper needs to know to do a good job with bamboo. This book is not intended to be a comprehensive source book. Instead our favorite 50 bamboos will be featured. They are organized two ways, first by height from shortest to tallest, and second by habit separated into two groups, those suitable for use in a hedge and those better suited for use as a specimen. Where the case study section featured ways to use a particular bamboo such as privacy hedges, this section will seek to reveal the strengths and weaknesses of each species featured.

Because we are organized by use, not name, if you have a mental image of the plant needed for a certain space, you should be able to find it fast. All of the species listed here are readily available.

When deciding what species to include the following question was asked: Does this species bring a unique and valuable feature to the landscaper's palette? If yes, it is included. Many have unique but not useful features and many are great bamboos but offer no advantage over another bamboo. They have been dropped. Just because they exist does not make them useful, so even though it is hard to leave anyone behind, we have done just that.

Another great advantage of this method is that the reduced number of plants allows a more in-depth study of each candidate, with more photos, identification tips and application ideas.

Each species description provides a plant height, diameter and minimum temperature that can be survived. Also provided are the leaf size and color and any unique identifying features.

There are many flaws to any simple classification of plant attributes. Plants just do not always follow the rules. In general I have listed more typical sizes, not the absolute largest on record. The size of many species will be dramatically affected by climate and sun.

An index at the end of the book provides a more normal alphabetical list of species included.

SMALL BAMBOO 1' - 4' TALL

White Stripe

(Dwarf White Stripe)

Pleioblastus fortunei

This shade-loving beauty will grow well indoors or out. It is a very popular house plant and a great place to get started with bamboo. The green-and-white variegated leaves look good year-round, will grow in partial sun or shade. It is very cold hardy as well.

- Height: 2 feet - 3 feet
- Diameter: 1/4 inches
- Minimum Temp: -10°F
- Shade or Partial Sun

Unique Identification Features

The short stature and long variegated leaves make white stripe easy to identify.

Landscape Details

The color of new leaves is extremely vibrant with a rich strong green contrasted against white. If grown in the shade, this color holds over the years but with full sun the green fades and the white yellows. Although still an attractive plant, we recommend cutting it all to the ground in the early spring. Within weeks the new shoots are back with vibrant colors again. You may use a weedwacker or lawn mower to cut it down.

White stripe spreads nicely and fills in in just a year or two. The rhizomes are small and shallow and it would be unusual to have any problem containing it with natural barriers such as walkways, drives, borders and such.

The only problem we have seen has been with Bermuda grass. Once fully infested, both use the same water, soil and sun, and unfortunately the Bermuda grass seems to be more invasive and will win the fight. Both must be removed and the white stripe replanted as all sprays we tried on the grass also killed the bamboo.

WHITE STRIPE

Above, white stripe is great in front of the fountain because it will not get much taller than a foot. It is a great complement to shiroshima behind the fountain which gets about 7 feet tall. To the left, white stripe fills in nicely under castillon and koi inversa in this planter. Below, this space has just been planted with 1- gallon plants and should fill in solid within two years.

Albostriata

(No common name)

Sasaella masamuneana 'Albostriata'

A very attractive low bamboo with wide cream-colored variegated leaves. Good in sun or shade as well as indoors.

- Height: 3 feet
- Diameter: 1/4 inches
- Minimum Temp: -10°F
- Full or Partial Sun

Unique Identification Features

- Leaves are glossy and thick
- Early leaves are more variegated that later leaves
- Leaves measure about 7 x 1.5 inches

Landscape Details

It is hard to think of a more perfect short bamboo. Its large flat leaves are stunning. It covers a large area well, makes a great bush when planted in a 2- foot ring of barrier and looks good in containers. Unlike white stripe, it holds its good looks even in full sun and remains tight and attractive in the shade.

ALBOSTRIATA
The shoot photo on the left will make plant identification easy. Notice above how the leaves layer near the edge of the planting and are more vertical in the center. Each leaf seems positioned to capture the most sun in this shady location.

SMALL BAMBOO 1' - 4' TALL

Green Stripe

(Dwarf Green Stripe)

Pleioblastus viridistriatus

Leaves are brilliant yellow with contrasting green stripes. May be cut down each year for more color. More sun turns leaves from yellow to white.

- Height: 3 feet
- Diameter: .25 inches
- Minimum Temp: 0° F
- Partial Shade Best

Unique Identification Features

- Green culms with a purple tint
- Banana-yellow leaves with random green stripes are quite distinct
- Leaves measure about 7 inches x 1 inch

GREEN STRIPE
The photo on the right shows new leaves with the yellow color but older leaves have bleached to white in this sunny location. Both lower right and left photos show green stripe in a shady location.

Landscape Details

Green stripe is indispensable when yellow is required. Like albostriata and white stripe, it is great in pots and as a shrub or ground cover. Unlike white stripe which fades out in the full sun, green stripe becomes more striking with the yellow bleaching to white and the green stripes remaining vivid.

Humilis

(No common name)

Pleioblastus humilis
also *Pleioblastus pumilus*
also *Pleioblastus gauntletti*
also *humilis var. pumilus*
also *P. argenteostriatus f. pumilus*

This aggressive ground cover is versatile and beautiful. Good in sun or shade, it makes an excellent erosion-control bamboo and its dark-green leaves contrast well with other variegated bamboo. Will grow in full sun but leaves look better in partial sun or shade.

- Height: 3 feet - 4 feet
- Diameter: .25 inches
- Minimum Temp: 0° F
- Sun or Partial Shade

Unique Identification Features

- Resources disagree about height, some saying humilis is taller (7') and pumilus shorter
- Leaves measure about 7 inches x .75 inches

Landscape Details

Humilis is a real workhorse in the landscape. It is a solid green all year long, with a soft rounded texture. As a hedge it may not get much taller than 2 feet but in a clump will see 3 or 4 feet. It fills in fast. It is also great in containers.

HUMILIS
A nice neutral green year-round makes this a fine shrub or hedge candidate. This one has lots of names and varies from nursery to nursery so be sure to check your source and look over a mature plant before committing to a truck load.

Veitchii

(Kuma-zasa)

Sasa veitchii

The large, dark green leaves turn white along the margins in the winter, giving the plant a unique and striking appearance, especially as a large clump viewed from a distance. This plant is very popular in the principal gardens of Japan.

- Height: 3 feet
- Diameter: .25 inches
- Minimum Temp: 0° F
- Likes shade

Unique Identification Features

- Margin on leaf is unique making this very easy to identify
- Leaves measure about 8 inches x 2 inches

Landscape Details

This bamboo is a must for any authentic Japanese garden. American customers vary in response so be sure they have seen it in winter.

In Japan it is typically used in large beds under hardwood trees. In the winter when the trees lack interest, the striking leaves keep the garden interesting.

The best way to keep it looking good all the time is to cut it all down in the early spring before the new shoots come up. Then it is green and fresh all summer and all the leaves develop the tan margins during the winter. Left to itself, those winter leaves will hang on another summer with extensive browning and be mixed with the fresh new leaves.

VEITCHII

This bamboo is not for everyone, but with a little care can be very impressive. The photos on the left are winter shots. Note that all the leaves have a uniform tan margin. The shot above is taken in the fall. The new leaves from this season are still fresh but last year's leaves are dying back dramatically. They will not last the winter. This mixed look can be avoided by cutting the whole stand to the ground in early spring.

Chino

(No common name)

Pleioblastus chino 'Angustifolia'
also Pleioblastus chino 'Elegantissimus'

An attractive bamboo from Japan with long, thin, weeping variegated leaves. A hardier alternative to Mexican weeping which is not very cold hardy. Good in sun or shade as well as indoors.

- Height: 6 feet
- Diameter: .5 inches
- Minimum Temp: 10° F
- Full Sun or Partial Shade

Unique Identification Features

- The long variegated leaves and weeping habit making this very easy to identify
- Leaves measure about 8 inches x 3/8 inches

Landscape Details

Mexican weeping bamboo gets a lot of press

because of its long thin weeping leaves. It is really nice looking at about 6 feet tall but it has two big problems. First it gets 20 feet tall over time and more importantly it is not very cold hardy. Chino is a great substitute in that it only gets 6 feet tall and is very cold hardy. The other difference is that chino is a pretty slow-running bamboo, but still should be contained to create that clump look. Chino is not particularly useful filling a large area as it becomes very dense, but is very good in pots, in a small ring of barrier or as a hedge.

CHINO
This bamboo is great in a clump or hedge. The clump below is contained with a ring of barrier and has not been trimmed in any way. The hedge on the right is perfect for this public garden.

Chinese Goddess

(Chinese Goddess Bamboo)

Bambusa multiplex 'Riviereorum'

Chinese goddess, a very slow-spreading clumping bamboo, has a low delicate fernlike foliage. Similar to fernleaf but a more stable and desirable form. It grows well in a container and in a protected or indoor location. This bamboo has the smallest leaves of any we feature.

- Height: 6 feet
- Diameter: .3 inches
- Minimum Temp: 12° F
- Sun or Partial Shade

Unique Identification Features

- Solid olive green culms
- Many short branches all growing upward on the arching culms
- Leaves measure about 1 inch x .2 inches in two rows

Landscape Details

Chinese goddess is very slow-growing and stable.

It is clumping bamboo that will form a perfect shrub with gracefully arching culms and tiny leaves. It forms a tight and dense clump with greater density over the years. It looks its best with some shade and when viewed from 10 to 20 feet away, but is very versatile.

CHINESE GODDESS

With a little trimming, a tight well-formed hedge can be made as seen on the right.   The tiny leaves run in two neat rows down each side of the short branches. A very attractive plant in the ground or in containers.

BAMBOO
GIANT

Fountain

(Fountain Bamboo)

Fargesia nitida

A mountain bamboo that enjoys a cooler climate and partial shade. It does well in coastal or mountain regions. It forms a fountain-shaped clump with fine dense leaves with sun and more open in the shade. Vertical, refined and tidy.

- Height: 10 feet
- Diameter: .5 inches
- Minimum Temp: -20º F
- Partial or Full Shade

Unique Identification Features

- Shoots late in season. Shoots rise above plant with no leaves until the next year.
- Shoots have a waxy blue-grey bloom
- Culms turn purple if exposed to the sun
- 5 to 18 branches per node
- More upright than other *Fargesia*
- Leaves measure about 2 inches x .6 inches

Landscape Details

This bamboo is the solution to many tough problems. It is hard to get good coverage but still keep a refined texture in those tough shady spots. Grows well directly under oaks, eucalyptus and redwood trees. This does well where nothing else does with the extremely tight clumping structure of its rhizomes.

FOUNTAIN
The planting above has been in this public garden for many years. The pot below is in a sunny spot in southern California and doing well. On the right, in the center of a redwood grove, not much else will grow but fountain is doing well. New shoots rise up tall by late summer (left) and remain with almost no leaves all winter. Next year they leaf fully and drop with the extra weight. A wonderful sight.

SMALL
BAMBOO
6' - 10'
TALL

Marmorea

(Marbled bamboo)

Chimonobambusa marmorea
Chimonobambusa marmorea 'Variegata'

The culms are supple, and range in color from light tan-green to deep red-maroon. The leaves are bright green and close together, almost bushy. It grows rapidly in good growing conditions. In autumn, culms turn red and the new shoots are covered by a marbled sheath. It can be used as a clump, hedge, or in a container. It originates from Japan. Form *variegata* is the same with variegated leaves.

- Height: 6 feet
- Diameter: .5 inches
- Minimum Temp: 5° F
- Full sun or partial shade

Unique Identification Features

- Dark red culms are very distinctive
- Tightly clustered leaves measure about 4 inches x 1/2 inches

Landscape Details

This is a fantastic bamboo with all kinds of interest. It fills in really fast and offers great texture and color. It is the perfect height to create privacy screens, shrubs or hedges. It becomes very dense so makes large areas inaccessible. Perfect for covering hillsides prone to erosion.

MARMOREA
Graceful arching culms with tight packed leaves make this one of the most beautiful bamboos. The nodes are quite pronounced and the initially green culms turn a dark red.

Tessellatus

(No common name)

Indocalamus tessellatus

Usually no more than 4 feet tall. It has the largest leaves of any bamboo in cultivation. Leaves have been seen up to 24 inches long and 4 inches wide but half that size is more typical.

- Height: 4 - 6 feet
- Diameter: .25 inches
- Minimum Temp: -5º F
- Shade Best

Unique Identification Features

- Brown persistent sheaths
- Long hanging leaves about 12 x 2 inches

Landscape Details

Although cold hardy, tessellatus gives the impression of being an exotic tropical plant. It is very effective in containers or planted as an understory, especially around water features. It and palmata have similar application but they are quite different. Palmata floats its leaves at waist level but this bamboo provides very high density right to the ground.

TESSELLATUS
This exotic tropical-looking bamboo is cold hardy and tough. On the right it is shown growing wild in China as an understory in a conifer forest. It is impressive contained in clumps or in pots as well as roaming free.

SMALL
BAMBOO
6' - 10'
TALL

Palmata

(No common name)

Sasa palmata f. nebulosa
also Sasa palmata

Palmata has large broad leaves, grouped in a "hand" shape at the end of the branch, almost seeming to float. They spread rapidly but do not become very dense.

- Height: 3 - 7 feet
- Diameter: .25 inches
- Minimum Temp: -15° F
- Shade or Partial Shade

Unique Identification Features

- Dense black spots on older culms
- Shoots start out lime green with white bloom
- Leaves cluster at top of culm with palm-shaped form
- Leaves thick-veined and tessellated
- Leaves measure about 12 inches x 4 inches

Landscape Details

Palmata is a tropical-looking plant like tessellatus but the leaves are in clusters at the end of each branch and there are generally few leaves on the lower part of the plant. Because they naturally spread out they form a very interesting understory with plenty of visual interest but open space as well.

Palmata spreads rapidly so containment is advised.

PALMATA
Exotic and tropical-looking but also a bamboo with a unique and distinct form. Clusters of leaves float at the end of almost vertical branches.

Shiroshima

(No Common Name)

Hibanobambusa tranquillans 'Shiroshima'

Strongly variegated leaves that persist all through the year. Forms thick hedge, good in many conditions.

- Height: 14 feet
- Diameter: 1.25 inches
- Minimum Temp: 0° F
- Sun or Partial Shade

Unique Identification Features

- Bright -colored blade
- Purple tone in new leaves - variable variegation
- Grooved internode
- Single branch per node, later developing into 2 or 3
- Leaves measure about 10 x 2 inches

Landscape Details

I saved shiroshima for last in this section because it is such an amazingly useful plant. In shade the colors are rich and in full sun, the canes darken and the contrast is striking. This plant is an accidental cross breed between veitchii and black bamboo, when both flowered in Japan at the same time, so it is literally a one-of-a-kind plant.

SHIROSHIMA

This very specialized plant initially has just one branch per node, resulting in the open form seen above. This is an installation for the TV show "Extreme Makeover". The location right along the entry gate will allow the color to be enjoyed. It is upright and extremely well -behaved. Over many years the form fills in as seen to the right and a stable size and color result. Although shiroshima spreads, not rampantly, so containment is quite easy. The color holds year-round and little maintenance is needed.

MEDIUM HEDGE BAMBOO 12' - 25' TALL

Arrow

(Arrow Bamboo)
Pseudosasa japonica

(Green Onion Bamboo)
Pseudosasa japonica 'Tsutsumiana'

Culms of arrow bamboo are straight and smooth and are used for many craft applications. The plant grows erect and only moderately spreads. Leaves are large and attractive. Will grow nicely in full sun although leaves look better with some shade, but it tends to get a little lanky in full shade. It is a favorite in containers or pots. It is wind and salt tolerant, and good in a wide range of soils.

Green onion bamboo is the same but with unique swollen internodes.

- Height: 8 feet - 15 feet
- Diameter: .5 inches
- Minimum Temp: 0° F
- Full Sun or Partial Shade

Unique Identification Features

- Persistent culm sheaths
- Single branch on upper culm followed by two smaller
- Long drooping dusky-green leaves measure about 8 inches x 1 inches

Landscape Details

Arrow is a great hedge bamboo which is very different from the more familiar timber bamboo with the larger culms. The culms are small and numerous and make an impenetrable barrier. It is tolerant of all kinds of abuse and difficult conditions.

ARROW and GREEN ONION
Arrow canes are straight as an arrow. The swollen nodes of green onion (below right) are unique and fun, but hardly show under the profusion of leaves and branches. A hedge of 15-gallon arrow plants (lower left) show how effective this plant is for privacy. From a distance the arrow hedge above is very neutral and just the right height to block one story without blocking the light or view.

ARROW

Arrow spells privacy. On the right, immediately after planting it already provides good coverage. Directly below, after a few years the coverage is total and the hedge is impenetrable.

MEDIUM

HEDGE

BAMBOO

12' - 25'

TALL

Linearis

Pleioblastus linearis

This dense and slightly arching bamboo is slow to grow and spread but good in full sun or shade. Makes a great hedge where more control is needed. The very attractive long dense grass-like leaves are elegant, bold and tidy. It is salt and wind tolerant.

- Height: 13 feet
- Diameter: 1 inches
- Minimum Temp: 0° F
- Sun or Shade

Unique Identification Features

- Shoots late
- Pale green culm color - new culms very pale
- Persistent green sheaths paling to buff with no hair or setae (bristles)
- Branches are very vertical as well as upper leaves
- Similar to *Pleoblastus gramineus* which has setae (bristles)
- Upright stiff leaves measure about 8 x .5 inches

Landscape Details

This is not a typical bamboo and is really looking for that unique place where total privacy is required, but the sharp angular clustering of the leaves will provide the needed texture.   The culms are small and numerous and make an impenetrable barrier.

LINEARIS
If its strength is its perfectly upright form (no bamboo is better-behaved) then its weakness is that it is slow to spread.

MEDIUM

HEDGE

BAMBOO

12' - 25'

TALL

Variegated Simonii

(Medake)
Pleioblastus simonii 'Variegatus'

Pleioblastus hindsii is a good alternate with similar form

This is a dense and upright hedge bamboo that is fast-spreading and quite drought-tolerant. It shoots late in the season. Some of the long narrow leaves are variegated and some are green. This bamboo is particularly good for screens and windbreaks. It is salt-tolerant.

- Height: 13 feet - 16 feet
- Diameter: 1 inches
- Minimum Temp: -10° F
- Sun or Partial Shade

Unique Identification Features

- Shoots often - even late in the season
- Shoots start out light green
- Persistent sheaths on culms and branches
- Some leaves variegated, typically much narrower - *Hindsii* leaves have unusual extended taper to fine point and no variegation
- Leaves measure about 8 x .75 inches

SIMONII and *HINDSII*

These bamboos look similar. *Hindsii* (top left) has an extended point on its leaves where *simonii* leaves (lower left) may sometimes be variegated. The sheaths are pale green as seen on the *hindsii* shoot (far left) and the sheaths are persistent.

Landscape Details

Although only a moderately beautiful bamboo, it is extremely effective as a privacy hedge, even if planted in pots. The persistent sheaths and somewhat tattered look to the leaves and branches lend it a very informal look but from a distance is very neutral and attractive.

MEDIUM HEDGE BAMBOO 12' - 25' TALL

Mountain Bamboo

Yushania anceps

This very vigorous clumping bamboo spreads rapidly like a running bamboo. It has performed well in containers, full shade and full sun. Its vertical culms arch near the tops, and masses of leaves forming at each node create a layered plume-like effect. A great hedge bamboo with tremendous architectural value, especially when viewed from a distance.

- Height: 13 feet
- Diameter: 1 inch
- Minimum Temp: 0° F
- Sun or Partial Shade

Unique Identification Features

- Very unusual long smooth rhizome neck (up to 6 feet long)
- Many branches cluster at the nodes
- Leaves measure about 5 inches x .5 inches

Landscape Details

Enough good things cannot be said about this clumping bamboo and the only bad thing is that it spreads more rapidly than most running bamboo. It will fill in a hedge or container in just one season. In the following years the culms are weighted down with an abundance of very attractive leaves. The culms are short so the arching does not become a problem even if planted near a walkway. The best way to describe it is FUN. It will lighten up the landscape and leave you smiling for sure.

MOUNTAIN
When you see this clumping bamboo spread you cannot help smiling at the simple running vs. clumping wisdom. The long rhizome necks (far left) have been measured at over 6 feet long. It is superb in hedges or containers.

MEDIUM

HEDGE

BAMBOO

12' - 25'

TALL

Yamadorii

(No Common Name)

Semiarundinaria yamadorii

This bamboo is of the same family as temple bamboo but more dense and more arching. It is very fast-growing and the canes display a range of color depending on contact with the sun, including green, yellow and red. It is quite rare.

- Height: 25 feet
- Diameter: 1.5 inches
- Minimum Temp: -5° F
- Full Sun Best

Unique Identification Features

- Culm colors in the sun
- Many branches cluster at the nodes
- Yellow-green leaves measure about 5 inches x 3/4 inch, with new shoots having much larger leaves

Landscape Details

Yamadorii grows vigorously in all conditions from full sun to full shade. In the shade, the color is light green throughout (leaves and culms). In full sun the culms show great color but the leaves tend more toward yellow-green.

Containment is required and some space for the occasionally leaning culms. A very dramatic and commanding bamboo.

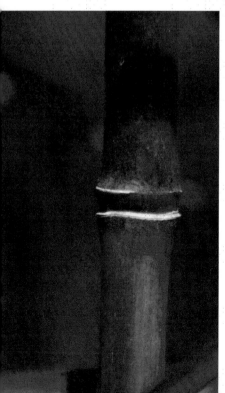

YAMADORII
Note the new culm is olive green (above) but with some sun changes to red and yellow. The plants on the right are a 2000-square foot grove started with one 25-gallon plant 5 years ago. I have not seen a mature planting of this rare bamboo to know about ultimate height and form.

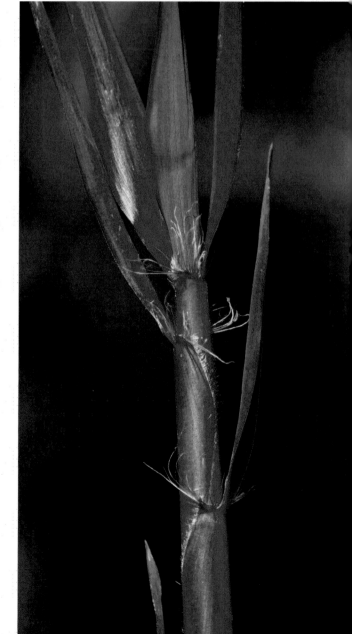

Square

(Square)

Chimonobambusa quadrangularis

Culms are square in cross-section. Square has a very columnar stature with supple, drooping leaves. It is great in full or partial shade and makes an impressive single clump. It is often used indoors in a pot or planter and will grow in full sun but leaves look better in partial sun or shade. An absolute must for a Japanese garden.

- Height: 12 feet - 15 feet
- Diameter: 1 inch
- Minimum Temp: 10^0 F
- Shade or Partial Sun

Unique Identification Features

- Square cross-section
- Aerial roots on nodes
- Spring and Fall shoots - good to eat
- Grey-green matt colored culm
- Leaves measure about 6 x .6 inches

Landscape Details

Square bamboo is a favorite in gardens throughout Japan. It is one of only two bamboo found in the Koishikawa Korakuen Garden of Tokyo. It is just the right height and density for screening and privacy. It seems to really love growing in a container.

SQUARE
Notice the tall columnar form even in this 15-gallon plant (left). It is great indoors and will handle a range of conditions including full shade. Leaf tips tend to burn easily giving it a casual look. I find a place for square in every project.

MEDIUM

HEDGE

BAMBOO

12' - 25'

TALL

Water Bamboo

(Water Bamboo)

Phyllostachys heteroclada 'Purpurata'

This unique bamboo has air channels in its rhizome so it can survive very wet conditions. The slender culms have a zigzag form and may lean or bend when wet.

- Height: 16 feet
- Diameter: .75 inches
- Minimum Temp: 0° F
- Full Sun Best

Unique Identification Features

- Slight zigzag form to slender culms
- Hollow rhizome
- Purple sheath blade and purple stained sheath
- Leaves measure about 3 inches x .5 inches

Landscape Details

Water bamboo looks like a typical medium-sized hedge bamboo. There many other species from the same *Phyllostachys* family that are a bit more interesting, but this bamboo has two features which warrant its inclusion in this species list. The zigzag habit of the culm is very interesting visually and perhaps compensates for the somewhat excessive arching of the culms under load. But the vital feature is the ability to handle damp ground better than any other bamboo. For that it is indispensable.

WATER BAMBOO
The key to the success of this bamboo in wet ground is its ability to deliver air to the roots. The rhizome is hollow and the individual roots shown in cross-section below have large air channels.

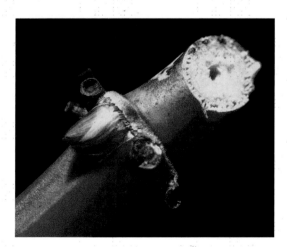

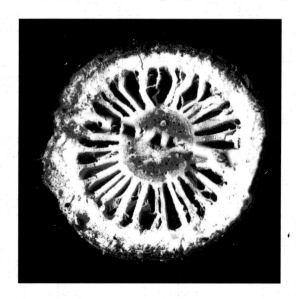

WATER BAMBOO
The plant below was growing in normal ground. Since moving it into a soggy place, notice the much larger darker shoots (far left). It is very happy in this place where other bamboos looked very poor. Notice the attractive zigzag of the lender culms. (left)

MEDIUM

HEDGE

BAMBOO

12' - 25'

TALL

Golden Bamboo

(Golden bamboo)

Phyllostachys aurea

The most commonly-cultivated bamboo in North America is golden bamboo. It is easily identified by one of several short internodes at the base of some of its canes. It is hardy and drought-tolerant and a very good bamboo for use as a potted plant. It is extremely aggressive - has given bamboo a bad name - and is not recommended without containment. In full sun it will yellow but stays green in partial shade or indoors.

- Height: 25 feet
- Diameter: 2 inches
- Minimum Temp: 0° F
- Grows best in sun but looks better in shade

Unique Identification Features

- Compressed internodes
- Leaves measure about 4 inches x .75 inches

GOLDEN
This bamboo gives bamboo its bad name. It is very aggressive. Notice the rhizomes shooting out of the chips next to the hedge on the right. Spring or fall, it is always moving. It is usually very dense, but given space as in the natural grove above it will open up and make a pleasant cover. The canes also make fantastic walking sticks. The older canes on the left have yellowed with exposure to the sun.

Landscape Details

Golden bamboo is the least attractive of the hedge bamboo in my opinion. It spreads so aggressively that all its energy is consumed and yellowing leaves are a very common result. I only recommend this bamboo in three cases:

- In containers likely to see very abusive conditions (this bamboo is almost indestructible) or when extremely fast full coverage is needed.
- In containers in a protected place where the canes can be exposed or when canes will be harvested for craft projects.
- This is a great bamboo for manipulation such as trimming or forced shaping. You can plant it in a shallow pot, trim off all the branches up to some point and then leave a tuft of green on top. Great fun and you're unlikely to hurt it.

We do also install it in the ground with containment, but only if conditions are likely to be harsh.

Hedge Bamboo

Phyllostachys rubromarginata
Equivalent substitutes also include:
Phyllostachys nuda
Phyllostachys propinqua
Phyllostachys dulcis

These bamboo are very generic green medium height bamboo without noticeable special features. They are the workhorses of the hedge business. Without detailed study of the shoots, you would be hard pressed to tell one species from the next. They are all upright, well behaved, all green and great for hedges. Nuda culms are a little darker green and it is the cold hardiest.

- Height: 15 feet - 25 feet
- Diameter: 1 1/2 inches
- Minimum Temp: -5 to -20°F
- Full or Partial Sun

Unique Identification Features

- *Rubromarginata* has red edging on the sheath and *dulcis* has spots
- Nuda is darker and has lots of white powder on the new shoots
- Leaves measure about 4 inches x .75 inches.

Landscape Details

These bamboos are easy to use, spread only moderately and look the way one expects bamboo to look. They make a great hedge that is not too tall, typically 20 feet.

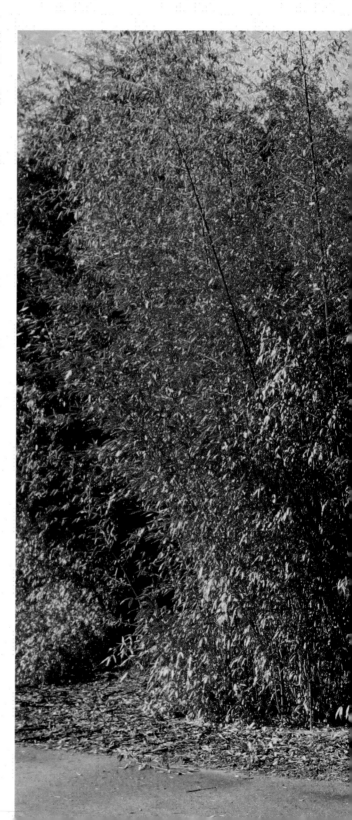

HEDGE

The shoots shown are quite different even though the bamboo looks the same. From left to right, *Ph. rubromarginata, Ph. dulcis and Ph. propinqua*. The *Ph. rubromarginata* hedge below is three years old and has never been trimmed, thinned or maintained. It is only 3 feet wide and provides 100% privacy right to the ground. If placed in front of a 6-foot fence, the lower branches could be removed to expose the culms.

Spectabilis and Yellow Groove

Phyllostachys aureosulcata 'Spectabilis'

(Yellow Groove)
Phyllostachys aureosulcata
 also similar to yellow groove is
Phyllostachys aureosulcata 'Alata'

These are the same bamboos with the colors inverted. Spectabilis is yellow with a green stripe and yellow groove is green with a yellow stripe. The internodes are rough like fish scales to the touch - smooth down and rough up. They have beautiful new shoots in the spring. Some individual canes zigzag near the base. A very hardy bamboo and very beautiful. Great for hedges, even in narrow spaces. Alata is the same as yellow groove but all green and straight.

- Height: 25 feet
- Diameter: 1.5 inches
- Minimum Temp: -10° F
- Full Sun Best

Unique Identification Features

- These bamboo are most easily identified by the unique roughness of the culm
- New shoots are very distinctive
- Occasional crooks in culm
- Leaves measure about 4 inches x .75 inches

Landscape Details

These are wonderful bamboo, with just the right form for hedges. They make a great hedge that is not too tall, typically 20 feet and are very upright and cold hardy. The inverse coloring is a great asset because they are the same bamboo with very different color. This means that the color of the leaves will be the same, the form is the same and the spreading and height will be the same. So both colors can be mixed in the same planting or spread around with fantastic results. These two are my favorite medium hedge bamboos.

SPECTABILIS
Shown to the right and immediate left, spectabilis has an attractive yellow color with a regular green stripe.

YELLOW GROOVE
Yellow Groove above and to the far left looks just like spectabilis but with the colors inverted, green with a pale yellow stripe.

Both species have random crooks in the occasional cane and both have very colorful shoots.

Borinda

Borinda angustissima

This fast-growing and striking clumping bamboo quickly becomes a focal point in any garden. It has very narrow small leaves, purple and green arching culms and impressive shoots. Great in a sheltered courtyard or next to a building in cold and windy areas.

- Height: 18 feet
- Diameter: 0.5 inches
- Minimum Temp: 15° F
- Partial Shade

Unique Identification Features

- Shoots have very long sheath covers initially purple or purple-green
- New shoots are purple and covered with a powdery white bloom.
- Leaves measure about 2 x .2 inches

Landscape Details

When placed in the sun, this beauty seems to be more compact in size, with a lower density of leaves and thus less weeping. It makes a great shrub which is approximately round. In a shady location it will reach full size with incredible results. The leaf density is higher in the shade with greater height and more weeping.

BORINDA

Such a profusion of perfectly patterned leaves is inspiring. I love borinda as a focal point in any space that cries out for peace of mind and tranquility. No hot tub or secret garden should be without it. This is a superior plant for a space of about 6 to 8 feet in diameter. With its weeping culms and enough light, height is rarely above 8 feet. In heavy shade it may go twice as high.

Robusta

Fargesia robusta

This very vigorous clumping bamboo creates a colorful addition to any garden. It has performed well in containers, full shade and full sun. Its many shoots are good to eat, and it is one of the bamboos preferred by pandas.

- Height: 10 - 20 feet
- Diameter: 1 inch
- Minimum Temp: 0° F
- Sun or Partial Shade

Unique Identification Features

- Shoots very early in spring
- Shoots start out lime green and crimson
- Sheaths quickly become matt white with dark hair
- Short branches aiming up at 45°
- Leaves measure about 5 x .75 inches

Landscape Details

Its upright form makes it available for tighter spaces than similar species which have more arching culms. The quantity of shoots it generates is very impressive so be careful to plan room for expansion or to contain. This is a very usable bamboo. I have not seen it much over 10 feet but I always seem to be splitting my clumps as they grow so fast. Wow!

ROBUSTA
Robusta is as its name implies, a very tolerant fast-growing clumping bamboo. The clump on the right doubled in size each year. Every other year it has been removed and quartered, with a quarter repla■ There are more shoots than culms in the photo above.

MEDIUM

SPECIMEN

BAMBOO

12' - 25'

TALL

Tibetan Princess

(Tibetan princess)

Himalayacalamus asper
also known as
Neomicrocalamus microphyllus

The normally green culms turn red in the sun. New shoots start out standing up (below) but fall the next season with the weight of leaves.

This mountain bamboo from China has attractive arching culms packed with dense leaves at each node, it is attractive around water features, along fences or in containers. When hit by the sun, the canes turn a brilliant red.

- Height: 15 feet
- Diameter: 1 inch
- Minimum Temp: -5 F
- Sun or Shade

Unique Identification Features to All *Himalayacalamus*

- About 15 branches per node first year, up to 40 over life
- Sheath is smooth on inside and upper part next to a short ligule
- Leaves measure about 5 x .5 inches and are not tessellated
- New shoots are good to eat. May shoot in fall and not leaf until spring

TIBETAN PRINCESS
The luxuriance of the foliage on the right is encouraged by partial shade. In full sun to the left, the plant is more compact.

Specific Unique Identification Features

- Sheath is rough and slightly hairy
- Blue powder common on new culms
- Culms turn very red in full sun.

Landscape Details

This is my favorite medium specimen bamboo. I love the color and form, especially near water features. The size can be a problem, as you should leave at least a 12- to 15-foot circle clear for this bamboo over time because of the arching culms. The *Himalayacalamus* bamboos are often mis-named so be careful with sourcing.

MEDIUM SPECIMEN BAMBOO 12' - 25' TALL

Mexican Weeping

(Mexican weeping)

Otatea acuminata ssp. aztecorum

This graceful weeping clumping bamboo spreads moderately. Its light green leaves will eventually obscure the culms completely. In temperate climates it should be in a sheltered warm location. In cooler locations it will be much smaller.

- Height: 20 feet
- Diameter: 1.5 inches
- Minimum Temp: 27° F
- Sun or Partial Shade

Unique Identification Features

- Clumping bamboo with long necks up to 2 feet
- Leaves measure about 5 x .75 inches

Landscape Details

A graceful and attractive bamboo, it is very useful in warmer climates. It is really not cold hardy enough for northern California except in atriums and warmer sheltered places, but this varies from source to source so be very careful to only buy plants from sources which experience colder temperatures than you do. Also be careful to plan for the right-size plant based on the climate you are working in. A more suitable substitute for Mexican weeping in a warm place when a small plant is desired is chino. Robusta, borinda or Tibetan princess are more attractive substitutes if the larger size is needed in a colder area.

MEXICAN WEEPING
In northern California our Mexican weeping stays small. The clump on the right has been here 10 years and is 4 feet tall and very attractive, although I lost 3/4 of it last winter during a cold snap. Shown above in southern California it is a monster.

South African

(South African Mountain Bamboo)

Thamnocalamus tessellatus

The only bamboo native to South Africa, it forms an open clump with new shoots traveling about 1 foot each year. Upright, attractive and quite drought-tolerant.

- Height: 14 feet
- Diameter: 1 inch
- Minimum Temp: 0° F
- Sun or Partial Shade

Unique Identification Features

- About 7 branches per node first year, up to 12 over life
- Persistent sheath is pale, almost white with erect blade
- Leaves measure about 3 x .4 inches and are tessellated
- Branch buds are taller than wide
- Branches rise vertically for about .5 inches then out at 45° angle
- Branches primarily on upper nodes
- Culms matt green sometimes aging to dull purple

Landscape Details

Although this bamboo can look a little scruffy because of the persistent culm sheaths, it makes a striking architectural form when mature, with a

lower section of straight culms and tight dense clump of foliage on top, almost mushroom-shaped. It strikes me as sharp, strong and male in character.

SOUTH AFRICAN
The sheaths hang on for the first year on the new shoots
giving a mixture of solid green culms and tan and green
striped culms.

Chusquea

(South American Mountain Bamboo)

Chusquea gigantea
also *Chusquea culeou*

This clumping mountain bamboo is from Chile. A unique form well-suited to coastal California. Tall, arching culms with tight branch groups. There are many chusquea species and much variation within the species.

- Height: 35 feet
- Diameter: 2 inches
- Minimum Temp: 0° F
- Sun or Partial Shade

Unique Identification Features

- Solid culm
- One dominant branch per node
- Very open clumper - necks up to 12 inches long
- Persistent sheaths
- Deep rhizome
- Leaves measure 5 x .4 inches
- Somewhat rare

Landscape Details

There is no mistaking this bamboo for a typical timber bamboo. The culms have a unique

appearance with one large branch and many small ones at each node. They form tufts that really show up on the otherwise quite straight and well-behaved culms. A clump of this bamboo spreads from 16 to 24 inches in diameter each

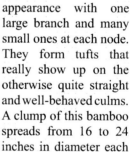

CHUSQUEA
The shoots of this bamboo are probably the most colorful of any bamboo listed. They shoot much of the summer and fall offering great color and fast-moving action.

MEDIUM

SPECIMEN

BAMBOO

12' - 25'

TALL

Aurea Koi and Inversa

(no common name)

Phyllostachys aurea 'Koi'
Phyllostachys aurea 'Flavescens-inversa'

Similar form to golden but much less aggressive. Unusual matched species with inverted colors. Koi canes turn from green to yellow after the first 6 months, but the cane grooves remain green. The inversa form is the opposite. The canes remain green but the grooves turn yellow. Upright and very tight-clumping.

- Height: 15 feet
- Diameter: 1.25 inches
- Minimum Temp: 0° F
- Full Sun Best

Unique Identification Features

- Compressed internodes at base of culm
- Koi is yellow with green stripe
- Koi inversa is green with yellow stripe
- Leaves measure 5 x .4 inches

Landscape Details

This bamboo is too slow-spreading to make a privacy hedge in any reasonable time. Instead it is very attractive in small clumps or in containers. The wonderful inverted nature of the two species is ideal for mixing. It should have all the lower branches removed to show off the striking culms. It is upright and can be very nicely shaped.

KOI and KOI INVERSA
These very showy bamboo have attractive canes but mediocre leaves, similar to golden with a tendency to yellow, especially in full sun. Notice the compression of the internodes occurs on the bottom of the culm (left) but the colors continue (upper left).

Himalayan Blue

(Blue Bamboo)

Himalayacalamus hookerianus

This is a mountain bamboo from China. In full shade, culms maintain blue coating. Some forms get larger than others. Very nice in containers, around water features and in full shade. Not very cold hardy.

- Height: 25 feet
- Diameter: 2 inches
- Minimum Temp: 25° F
- Sun or Partial or Full Shade

Unique Identification Features to all Himalayacalamus

- About 15 branches per node first year, up to 40 over life
- Sheath is smooth on inside and upper part next to a short ligule
- Leaves measure about 5 x .5 inches and are not tessellated
- New shoots are good to eat. May shoot in fall and not leaf until spring

Specific Unique Identification Features

- Sheath is long, with long ligule and long reflexed blade
- Blue powder on new culms
- Culms can age to yellow - green or purple

Landscape Details

This is an attractive bamboo but the "blue" in the name is deceiving. New shoots start out with a bluish haze on the surface which fades almost immediately except in full shade. Ignoring the blue issue, it is a tall upright majestic bamboo with very attractive leaves. Unfortunately it is not very cold tolerant and most of what was growing in my backyard died to the ground last winter.

BLUE
The culms start out with blue haze (far left) but fade quickly to green (above) with some sun and to yellow (left) with full sun. The leaves are large and extremely graceful, a real strength of this bamboo as well as its very vertical structure, even in full shade.

MEDIUM

SPECIMEN

BAMBOO

12' - 25'

TALL

Candystripe and Silverstripe

(Candystripe)
Himalayacalamus falconeri 'Damarapa' also known as
Drepanostachyum hookerianum

(Silverstripe)
Bambusa multiplex 'Silverstripe'

Although these two bamboos have very different taxonomy, I find that they fit into the same place in the landscape. Candystripe is a mountain bamboo from China which has green culms striped with yellow or pink.. Silverstripe is from the subtropical genus *Bambusa* but unlike most of the larger timber bamboo in this genus has very attractive leaves. The culms have occasional white stripes and the leaves are sharply variegated. Both have attractive arching culms packed with dense leaves at each node, and are attractive around water features, along fences or in containers.

- Height: 15 - 20 feet
- Diameter: 1 inch
- Minimum Temp: 15° F
- Sun or Partial Shade

CANDYSTRIPE
Candystripe is an upright clumper with small attractive leaves (above). Notice the range of form (both lower right). The culms are striped with yellow or pink (below).

SILVERSTRIPE
Silver-stripe is also an upright clumper with small variegated leaves (upper right and lower left) and white striped culms.

Unique Identification Features

- Silverstripe has random white stripes on culm and leaves
- Candystripe has green culms with yellow or pink stripes
- Leaves measure about 2.5 x .3 inches

Landscape Details

These bamboo are clumping bamboo and still exhibit the V-shape form, with tightly clustered culms at the bottom spreading at the crown, but less so than the many clumpers earlier in this section. They look more like the clumping timber bamboo described later but without the sloppy disheveled leaves and excessive height.

Alphonse Karr

(No common name)

Bambusa multiplex 'Alphonse Karr'

A common clumping bamboo known for the yellow culms with irregular green stripes. Culms get a little red in the sun. Growth will be slow in areas with a cool summer. Makes a good houseplant.

- Height: 20 feet (vary much with conditions)
- Diameter: 1 inch
- Minimum Temp: 20° F
- Sun or Partial Shade

Unique Identification Features

- The yellow culm with random green stripes are distinctive
- Leaves measure about 4 inches x 1 inch

Landscape Details

This is a very common bamboo but its lack of cold tolerance makes its size and growth rate dependent on the local conditions

ALPHONSE KARR
The clump to the right is growing in southern California and is full sized. Smaller sizes can be expected where summers are cool. Some pretty amazing 5-gallon plants can be produced (below left). Notice the red staining on the left caused by the sun on these new culms. Although a little scruffy with the persistent culm sheaths, it is certainly a neat attractive addition to any landscaping project.

Robert Young

(no common name)

Phyllostachys viridis 'Robert Young'
also *Phyllostachys viridis 'Houzeau'*

Robert Young canes turn sulphur-green to a bright banana-yellow with bright green stripes of variable width displayed on many internodes. A truly exotic, tropical-looking timber bamboo which is very hardy. It likes clay-type soils and is good as an isolated clump or grove. Very slow-spreading in northern California so dense clumps are formed. Basic form is *Viridis* which is all green and faster spreading. Another form is *Houzeau* which is green with yellow stripe in groove.

- Height: 30 feet
- Diameter: 3 inches
- Minimum Temp: 0° F
- Full Sun Best

Unique Identification Features

- The yellow culm with random green stripes are distinctive
- Leaves measure about 4 inches x 1 inch

Landscape Details

In warmer locations this will make a great hedge bamboo as it is tall, has striking color and is generally upright and well-behaved. In cooler locations it may take an extra year or two to fill in completely. It makes a great specimen where a flash of yellow is needed but spreading cannot be tolerated.

ROBERT YOUNG
The grove on the right shows the high density of this slow-spreading runner. Shoots are initially green but change to yellow during the first year. The green stripes are random. The variety *Houzeau* is green with a yellow stripe in the groove (above right) but otherwise looks the same. *Viridis* is the parent of both these species and is all green. Its shoot (left) is typical of all three varieties. A distinct feature of these bamboo is a very sharp upright angle to the branches.

Leopard

(Leopard Bamboo or Snakeskin)

Phyllostachys nigra 'Bory'

One of the giant timber bamboo, its culms get blue-gray-green with random black splotches, never turning completely black. With beautiful, dark-green dense foliage it is very exotic but cold hardy. Good for isolated clumps, tall hedges and groves.

- Height: 40 feet
- Diameter: 4 inches
- Minimum Temp: 0° F
- Full Sun Best

Unique Identification Features

- Spots are unique
- Leaves measure about 4 inches x 1 inch

LEOPARD
Exotic, upright and moderately spreading. A perfect bamboo.

Landscape Details

Customers either love or hate this bamboo. It is either exotic or "looks sick". For those who love it, it can be used in many ways but most impressive is close at hand, where the branches can be stripped off to reveal the great culm patterns.

TALL

HEDGE

BAMBOO

25' - 40'

TALL

Black Groove

(No Common Name)

Phyllostachys nigra 'Megurochiku'

Similar to henon, but the culm grooves are brown or purplish-black. Beautiful, lacy green foliage. Quite rare. Makes a great hedge.

- Height: 35 feet
- Diameter: 3 inches
- Minimum Temp: 0° F
- Full Sun Best

Unique Identification Features

- Regular black stripe in groove
- Leaves measure about 4 inches x 1 inch

Landscape Details

This bamboo fits in everywhere. It is very upright and well-behaved. It is a very rich solid green color. The black stripes add a little interest without dominating. The leaves are small and lacy. This is an excellent bamboo but may be a little hard to find.

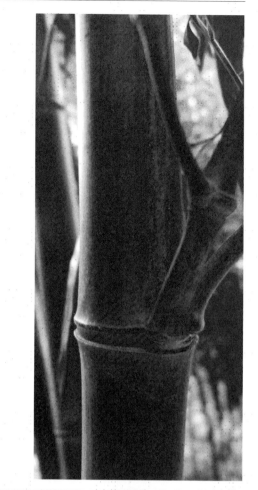

BLACK GROOVE
Although not a giant, tall enough for any privacy need, black groove is an elegant bamboo very fitting for even formal spaces.

Crookstem

(No common name)

Phyllostachys bambusoides 'Slender Crookstem'

A rare form of *Bambusoides* with irregular crooks in the canes. Medium-sized and exotic. Little is known about it.

- Height: 30 feet
- Diameter: 2 inches
- Minimum Temp: 0° F
- Full Sun

Unique Identification Features

- Smooth S-like bends
- Leaves measure about 4 inches x 1 inch

Landscape Details

This bamboo is very much like many other large green timber bamboo except for the occasional crooks. It is quite slow-spreading and well-behaved.

It may be quite difficult to find in quantity but worth the search for the right application.

CROOKSTEM
The crooks in this bamboo are not like yellow groove and spectabilis, where crooks develop when a node changes the angle of the culm like a hinge. The internode is straight but at an angle. Crookstem makes a smooth S-sweep. It is very graceful and unique among bamboo.

Violascens

(No common name)

Phyllostachys violascens

This is quite unusual because it goes through several color variations. Starts with dark, purple streaks which disappear in a few weeks. Gets yellow stripes changing to brownish-crimson stripes. Some culms remain green. Large leaves with blue/green insides. Very fast-growing. Makes a great hedge.

- Height: 30 feet
- Diameter: 3 inches
- Minimum Temp: 0^0 F
- Full Sun Best

Unique Identification Features

- Shoots early in the season
- Culm wall thin
- Colors very distinctive
- Leaves measure about 4 inches x 1 inch

TALL HEDGE BAMBOO 25' - 40' TALL

Landscape Details

Violascens is a fast and colorful bamboo. It is not that it spreads at such a great rate as golden bamboo does, but in the area where it does spread, it will produce large numbers of really big new shoots. A few 25-gallon plants will fill 400 feet in just one season! The only downside to this bamboo is that the thin wall culms are not as strong as other bamboo and can be more easily snapped off in a storm.

VIOLASCENS
The new culms are green, liberally streaked with purple. (left) As they age, the streaks fade to reddish-brown (right). In full sun the streaks become a darker brown as seen above.

Tanakae

(No Common Name)

Phyllostachys bambusoides 'Tanakae'

This is a very rare bamboo. It has an upright form and is slow-spreading. Recommended for collectors and as specimen.

- Height: 30 feet
- Diameter: 3 inches
- Minimum Temp: 0° F
- Full Sun or Partial Shade

Unique Identification Features

- Small brown spots increase in density each year
- Leaves measure about 4 inches x 1 inch

Landscape Details

This bamboo has particular value in that it tends to be quite large in diameter for its height. It is not very tall yet 2 1/2-inch culms are common.

TALL

HEDGE

BAMBOO

25' - 40'

TALL

TANAKAE
A common customer request is for a bamboo with really big culms but very short. This bamboo is as close to that impossible request as you can get. The internode length is quite short giving a compact, upright bamboo that provides impressive culms without excessive height.

TALL

HEDGE

BAMBOO

25' - 40'

TALL

Temple

(Narihira)

Semiarundinaria fastuosa
Semiarundinaria fastuosa 'Viridis'

Culms are straight and erect. Short branching results in a columnar, stately structure. Often used outside temples in Japan. Culms may turn from green to a purple/bronze color. It doesn't get any better than this.

- Height: 20 feet - 30 feet
- Diameter: 1 1/2 inches
- Minimum Temp: -5°F
- Full or Partial Sun

Unique Identification Features

- Many short branches
- Culms remain green in form 'Viridis' (below)
- Leaves measure about 4 inches x 1 inch

Landscape Details

Temple bamboo is very useful because of its short branching and columnar structure. It makes a very tall narrow hedge and can be used to make tall narrow clumps of exceptional beauty.

TEMPLE

New shoots and sheaths sport a rich purple color. Canes are a dark green fading to bronze in the sun. Tall and narrow form is quite unique and can be used in many difficult landscaping situations.

Giant Black

(Giant Black Timber)

Phyllostachys nigra 'Daikokuchiku'

This California species is much larger than black bamboo. Its large, jet-black canes turn black faster as well. New shoots are green but turn black by next season. Beautiful, lacy green foliage. A stunning giant. Good for grove and isolated clump but best mixed with other bamboo. Will grow in full sun but looks better as understory or with partial shade.

- Height: 20 feet - 40 feet
- Diameter: 2 inches
- Minimum Temp: 0° F
- Partial Sun

Unique Identification Features

- Colors very distinctive
- Leaves measure about 4 inches x 1 inch

Landscape Details

Black is a very popular bamboo and giant black is stellar. Black bamboo does present several challenges. The density of foliage is quite low so even a hedge 6- or 8- feet thick may give only

partial privacy. Black culms also tend to lean unpredictably which can be awkward in certain locations. It is best mixed into a hedge or grove of another color or placed in strategic locations where close proximity and handling of the culms is likely, and filtered light is desirable. I have black bamboo right outside my dining room window so I can see the canes each day, but not have my view or the

GIANT BLACK
New culms come up green (lower left) and change to black within a year. The second-year cane on the right is just finishing the transition and will be jet black by the time the year is out. As beautiful as black bamboo culms are to look at, the foliage is thin and sparse as seen in the 36-inch box (center below). A larger grove can be very impressive (below right).

TALL
SPECIMEN
BAMBOO
25' - 40'
TALL

Castillon and Inversa

(no common names)

Phyllostachys bambusoides 'Castillon'
Phyllostachys bambusoides 'Castillon Inversa'
Phyllostachys bambusoides 'Allgold'

Castillon and allgold, another form, have the most brilliant gold yellow canes of any bamboo. Castillon has contrasting green grooves. Castillon inversa, another form, is green with contrasting yellow grooves.

All have dark green leaves with an occasional yellow stripe making them very showy. They have long branches and the canes tend to lean so it is better for single-clump groupings, grove or potted plant than in a narrow hedge.

- Height: 25 feet
- Diameter: 2 inches
- Minimum Temp: 5° F
- Full Sun Best

Unique Identification Features

- Colors very distinctive
- Leaves measure about 4 inches x 1 inch

Landscape Details

These are the most colorful bamboo and can be used to brighten any landscape. The bonus of having a matched but reversed color species

CASTILLON and INVERSA The inverted colors of castillon (left) and castillon inversa (below) are very attractive when planted together (right). Allgold (lower left) is brilliant gold and the leaves of each are variegated.

means that   mixed plantings can be accomplished. These are big bamboo and need plenty of room to spread out above the planting area. They are quite slow-spreading so cane density is pretty tight. Allgold has the most striking color but seems overly susceptible to aphid attack.

TALL

SPECIMEN

BAMBOO

25' - 40'

TALL

Buddha's Belly

(Buddha's Belly)

Bambusa tuldoides 'Ventricosa'

A suptropical genus which is fast-growing and very dense but not very cold hardy in northern California. This variety is know for a dwarfed condition which can occur in hot dry conditions in which the internodes are very short and swollen. In normal conditions, it is a very attractive ornamental bamboo with broadly arching culms. Good in a container and in a protected or indoor location.

- Height: 20 to 40 feet
- Diameter: 2 inches
- Minimum Temp: 15° F
- Sun or Partial Shade

Unique Identification Features

- Internode 8 to 14 inches long
- Three primary branches in each group
- Sheath with brittle unsymmetrical top, brown ring below
- Auricles of unequal size with one twice the other
- Wavy cilia on margin
- Leaves measure about 5 x .5 inches with short hair on back

Landscape Details

The smaller leaves of this clumper make it one of the most attractive of the large *Bambusa*. The broad arching culms do require some space but the effect is pleasing.

BUDDHA'S BELLY
The plant on the right and below right was forgotten in the spring watering schedule. It dried out and lost all its leaves. We started watering again and the summer sun turned the canes bright orange before it re-leafed. Its small new leaves and orange canes make an awesome sight in the early morning sun of this fall day. The canes are normally green with three main branches and a number of smaller ones shown below left.

Henon

(No Common Name)

Phyllostachys nigra 'Henon'

One of the giant timber bamboo. The distinctive-looking canes are slightly rough to the touch and blue/grey/green in color, with beautiful, lacy green foliage. Good in clay and quite drought tolerant. Makes a great hedge.

- Height: 40 feet
- Diameter: 4 inches
- Minimum Temp: 0^0 F
- Full Sun Best

Unique Identification Features

- Culms very slightly rough to touch
- Leaves measure about 4 inches x 1 inch

Landscape Details

Henon is the perfect choice for a tall hedge intended as a neutral green backdrop. It is beautiful and refined but discrete. It spreads quite slowly so is not a big problem to control.

GIANT HEDGE BAMBOO 40' - 60' TALL

HENON

Henon looks similar to many other bamboo especially when plants are small. The features of the shoot (far left) are distinctive and there is a slight roughness to the culm. On mature plants, the size, blue-green color and lacy foliage make henon easy to spot. It spreads pretty slowly so high culm density is seen.

Vivax and Green Stripe Vivax

(Vivax)

Phyllostachys vivax
Phyllostachys vivax 'Aureocaulis'

A fast-growing timber with shiny, bright green culms. Similar to *bambusoides* but considerably hardier and more attractive. Green stripe has the characteristics of vivax but with bright yellow culms with bold green stripes. Vivax has quite large bright green leaves. It stands upright and quickly forms a hedge or grove. Although uncommon, vivax is in great demand.

- Height: 35 feet - 50 feet
- Diameter: 5 inches
- Minimum Temp: -5° F
- Full or Partial Sun

Unique Identification Features

- Culms very shiny and smooth
- Leaves measure about 6 inches x 1 inch

Landscape Details

This is our all-round favorite timber bamboo. Its leaves hold their rich green color even when the

bamboo is shooting.   It is the dominant bamboo, taking control of every setting and stealing the show from every other species.   It will achieve the largest diameter more quickly than any other species in northern California.

VIVAX

The 5-inch culms soar 50 feet in the air creating a quiet meditative space below.   Where exposed to light (below) long well-organized leaves provide total privacy.   The green stripe version (below left) is a stunning option, incorporating all the benefits of vivax with a bold color as well.

Weaver's Bamboo and Oldhamii

(Weaver's Bamboo)
(Giant Timber Bamboo)

Bambusa textilis
Bambusa oldhamii

GIANT

SPECIMEN

BAMBOO

40' - 60'

TALL

Both these are a suptropical genus which is fast-growing and dense but not very cold hardy. Oldhamii is the most common but I think least suited to the task. Weaver's bamboo is a little more cold hardy and a little more attractive and might be a better choice where a giant clumping timber bamboo is needed. Weaver's bamboo is the most attractive and graceful of the giant *Bambusa*. It is strongly vertical with a small arching at the culm tips. The internodes are long and branching starts quite high. It is neater and more compact than oldhamii. Both are good in a container and in a protected or indoor location.

- Height: 40 feet (Oldhamii up to 60')
- Diameter: 2 - 4 inches
- Minimum Temp: 15°F (Oldhamii 20°F)
- Sun or Partial Shade

Unique Identification Features

- *Textilis* has a thin-wall culm with long internode and smooth nodes. Oldhamii has a thicker wall
- *Textilis* branches are of equal size. Oldhamii has one main branch with up to 30 smaller branches
- Lower nodes of *Textilis* are free of branches and branch buds
- *Textilis* leaves measure about 5 x 1 inch and are square-shaped and flat at base. Oldhamii leaves measure about 6 x 1.5 inches and are not tesselated

WEAVER"S and OLDHAMII BAMBOO
The leaves of these bamboo are large, floppy and look disorganized with oldhamii (far left) being the worst offender.

Landscape Details

It is rare indeed that I cannot find a better choice than the scruffy, overly dense clumping bamboo oldhamii. I include it here only because it is so commonly specified and used, more from ignorance of a better option than by preference, I believe. Weaver's bamboo meets the specifications with at least some redeeming qualities but in most cases I prefer to leave both these species behind.

Moso

(Moso or Giant Chinese Timber)

Phyllostachys edulis, formerly
Phyllostachys heterocycla pubescens

This is potentially the largest of the hardy timbers, with beautiful, giant, pale green canes that are soft and fuzzy, with contrasting small feathery leaves and arching tops and a terraced branch structure. Fresh moso shoots are a real delicacy in the Orient. Young shoots are covered with a fine, fuzzy covering that looks like velvet. This is a true giant. One of the most beautiful and probably the most magnificent of the giant hardies. It needs fertile soil, and some humidity to thrive.

- Height: 35 feet - 65 feet
- Diameter: 4 - 6 inches
- Minimum Temp: $0^{\circ}F$
- Full or Partial Sun

GIANT SPECIMEN BAMBOO 40' - 60' TALL

Unique Identification Features

- Fuzzy culms and hairy shoots make this bamboo easy to spot
- Leaves measure about 5 x 1 inch

Landscape Details

If oldhamii is the ugliest of the giant timbers, then moso is the most beautiful. It is a little harder to use than most bamboo, however. When planted in a restricted space, moso downsizes in height to fit the root space available so it does quite poorly in containers or narrow hedges. Plan at least 6 feet wide for a hedge, with 12 feet better. A real stunning bamboo to view from a distance, the layering of the branches gets better looking every year.

MOSO
Soft, delicate and lacy, yet able to dominate the mountainous region of China near Anji shown below. This bamboo makes a great backdrop as well as a wonderful open and accessible grove if you can spare the space.

Indexes

Salt and Wind Tolerant Species

Drought Tolerant Species

Saturated Ground Tolerant Species

Cold Tolerant Species

Special thanks to Harriet, above, who trekked around China showing me bamboo and feeding me bamboo shoots.

Below, a water feature at the Nursery including palmata, tessellatus, dwarf fernleaf and fountain bamboo against a backdrop of giant black.